KNOWLEDGE DRIVEN DEVELOPMENT

Public Food Policy and Global Development series
Edited by Suresh Chandra Babu

Providing expert insights from around the world into the key questions
for effective policy development.

For a full list of volumes in the series visit www.store.elsevier.com

KNOWLEDGE DRIVEN DEVELOPMENT

PRIVATE EXTENSION AND GLOBAL LESSONS

Edited by

YUAN ZHOU

Syngenta Foundation for Sustainable Agriculture, Basel, Switzerland

SURESH CHANDRA BABU

International Food Policy Research Institute, Washington, DC, USA

AMSTERDAM • BOSTON • HEIDELBERG • LONDON
NEW YORK • OXFORD • PARIS • SAN DIEGO
SAN FRANCISCO • SINGAPORE • SYDNEY • TOKYO
Academic Press is an imprint of Elsevier

Academic Press is an imprint of Elsevier
32 Jamestown Road, London NW1 7BY, UK
525 B Street, Suite 1800, San Diego, CA 92101-4495, USA
225 Wyman Street, Waltham, MA 02451, USA
The Boulevard, Langford Lane, Kidlington, Oxford OX5 1GB, UK

Notices

Knowledge and best practice in this field are constantly changing. As new research and experience broaden
our understanding, changes in research methods, professional practices, or medical treatment
may become necessary.

Practitioners and researchers must always rely on their own experience and knowledge in evaluating and using
any information, methods, compounds, or experiments described herein. In using such information or methods
they should be mindful of their own safety and the safety of others, including parties for whom they have a
professional responsibility.

To the fullest extent of the law, neither the Publisher nor the authors, contributors, or editors, assume any liability
for any injury and/or damage to persons or property as a matter of products liability, negligence or otherwise, or
from any use or operation of any methods, products, instructions, or ideas contained in the material herein.

ISBN: 978-0-12-802231-3

British Library Cataloguing-in-Publication Data
A catalogue record for this book is available from the British Library

Library of Congress Cataloging-in-Publication Data
A catalog record for this book is available from the Library of Congress

For Information on all Academic Press publications
visit our website at http://store.elsevier.com/

Working together
to grow libraries in
developing countries

www.elsevier.com • www.bookaid.org

Publisher: Shirley Decker-Lucke
Acquisition Editor: Nancy Maragioglio
Editorial Project Manager: Carrie L. Bolger & Billie Jean Fernandez
Production Project Manager: Lucía Pérez
Designer: Maria Inês Cruz

Typeset by MPS Limited, Chennai, India
www.adi-mps.com

CONTENTS

LIST OF CONTRIBUTORS

Kolawole Adebayo
Department of Agricultural Economics and Farm Management, Federal University of Agriculture, Abeokuta, Nigeria

Safiul Islam Afrad
Banga Bandhu Sheikh Mujibur Rahman Agricultural University, Gazipur, Bangladesh

Suresh Chandra Babu
International Food Policy Research Institute (IFPRI), Washington, DC, USA

Kristin Davis
Global Forum for Rural Advisory Services (GFRAS), Pretoria, South Africa; International Food Policy Research Institute (IFPRI), Washington, DC, USA

Tran Tri Dung
Centre for Creativity and Innovation, Boise State University, Boise, ID, USA; DHVP Research and Consultancy, Hanoi, Vietnam

Vinod Gupta
Krishi Vigyan Kendra, Sher-e-Kashmir University of Agricultural Sciences and Technology of Jammu, Jammu, India

G.D.S. Kumar
Directorate of Oilseeds Research, Rajendranagar, Telangana State, India

Rakesh Nanda
Division of Agricultural Extension Education, Sher-e-Kashmir University of Agricultural Sciences and Technology of Jammu, Jammu, India

Pham Hoang Ngan
Vietnam Inclusive Innovation Project, Hanoi, Vietnam

Natarajan Ramesh
International Food Policy Research Institute (IFPRI), Washington, DC, USA

S.V. Ramana Rao
Directorate of Oilseeds Research, Rajendranagar, Telangana State, India

Rahman Sanusi
Department of Agricultural Economics and Farm Management, Federal University of Agriculture, Abeokuta, Nigeria

Raj Saravanan
Agricultural Extension and Rural Sociology, Central Agricultural University, Pasighat, Arunachal Pradesh, India

Cristina Sette
Institutional Learning and Changes Initiative (ILAC), Bioversity International, Rome, Italy

Rakesh Sharma
Krishi Vigyan Kendra, Sher-e-Kashmir University of Agricultural Sciences and Technology of Jammu, Jammu, India

Caitlin Shaw
International Food Policy Research Institute (IFPRI), Washington, DC, USA

Motunrayo Sofola
Department of Agricultural Economics and Farm Management, Federal University of Agriculture, Abeokuta, Nigeria

Gaytri Tandon
Sarveshwar Organic Foods, Jammu, India

K.S. Varaprasad
Directorate of Oilseeds Research, Rajendranagar, Telangana State, India

Fatema Wadud
Directorate of Agricultural Marketing, Ministry of Agriculture, Dhaka, Bangladesh

Yuan Zhou
Syngenta Foundation for Sustainable Agriculture, Basel, Switzerland

ABOUT THE EDITORS

Yuan Zhou is the Head of Research and Policy Analysis at the Syngenta Foundation for Sustainable Agriculture, headquartered in Basel, Switzerland. She advises and supports the Foundation and its partners on policy development in agricultural extension, food security, biodiversity conservation, sustainable land and water management, and payment for ecosystem services. Before joining the Foundation, Yuan was a researcher at the Swiss Federal Institute of Aquatic Science and Technology (EAWAG), working on water and environmental policies, integrated analysis of water/food/environment relations, and rural development issues. Yuan holds a PhD in Environmental Economics from University of Hamburg in Germany and an MSc in Water and Environmental Resources Management from UNESCO-IHE Institute for Water Education in the Netherlands. She has published in academic journals on a range of topics related to environmental economics, farmer decision-making processes, agricultural extension, agricultural water management, and the economics of desalination and water transport.

Suresh Chandra Babu is a Senior Fellow and a Program Leader for the Capacity Strengthening Program at the International Food Policy Research Institute (IFPRI), Washington, DC. He has held several positions before joining IFPRI, including Research Economist, Cornell University, Ithaca, NY; Evaluation Economist, United Nations Children's Fund; Senior Lecturer, University of Lilongwe, Malawi. He has been a coordinator of the policy research program for Central Asia. He has held or currently holds visiting honorary professorships at the American University, Washington, DC; Indira Gandhi National Open University, India; University of Kwazulu-Natal, South Africa; and Zhejiang University, China. He currently serves on the editorial boards of several academic journals including *Food Security, Agricultural Economics Research Review, African Journal of Agricultural and Resource Economics, Journal of Sustainable Development, Food and Nutrition Bulletin,* and *African Journal of Food, Nutrition and Development.* He received his PhD and MS in Economics from Iowa State University, Ames, Iowa.

PREFACE

Agricultural transformation is closely linked to rising farm productivity. This process requires continuous adoption of new technologies, adjustment to changing institutions, and responses to government policies and programs designed to help farming communities. Agriculture is therefore heavily knowledge driven, in both developing and more developed countries. The main conduit for knowledge sharing is extension.

New demands on the public extension system include helping farmers cope with climate change and other threats to their natural resources such as soil and water. These challenges require extension officers to go beyond traditional technology transfer to other forms of knowledge sharing. That pressure generates innovations in the way they provide extension.

Service provision is, however, increasingly pluralized; the public sector is not the only source of extension. Companies that sell inputs or buy produce have a particular interest in two-way sharing of knowledge with farmers. Extension not only brings new knowledge to farmers, but also takes their challenges back to solution providers, innovators, and researchers.

Whichever extension model they choose, private enterprises either directly or indirectly charge farmers for the services. So far, though, observers have lacked a deeper understanding of the motivation, cost, benefits, efficiency, effectiveness, and sustainability of private-sector extension. This book aims to present a comprehensive assessment. It covers a wide range of agroecological conditions, crop choices, and institutional mechanisms worldwide. We present 10 case studies conducted using similar methodologies and seeking to answer similar questions in Africa, Asia, and Latin America. Although their insights are diverse, they point to a common set of challenges and solutions emerging across developing countries. Our findings are supplemented by a survey of the literature on private extension and a historical overview of its development. We suggest several resulting lessons for scaling up private extension models, and we offer guidance for policymakers intending to support the development of public–private partnerships in extension.

The idea for this book originated in discussions at the Beijing Roundtable in Agricultural Extension in Asia, held in March 2012. Organized by the Syngenta Foundation for Sustainable Agriculture, the event enabled extension experts and practitioners from four Asian countries to share experience and foster learning. International Food Policy Research Institute (IFPRI) contributed strongly to the Roundtable, and this collaborative book builds on the two organizations' longstanding interest in extension. The authors deeply appreciate the financial support of the

Syngenta Foundation for Sustainable Agriculture, and acknowledge the additional funding from IFPRI and the Global Forum for Rural Advisory Services. We are also very grateful for the support of a wide range of individuals and groups, without whom publication of this volume would not have been possible. These include the staff interviewed at various companies in order to develop the case studies, extension workers in the field, and the farming communities that shared their knowledge and gave feedback. In particular, the authors thank employees of Syngenta (Latin America North) working on the Frijol Nica program, as well as at Jain Irrigation, EID Parry, Marico, and Sarveshwar Organic Foods in India, PRAN in Bangladesh, AGPP Joint Stock Company in Vietnam, Kenya Horticultural Exporters in Kenya, Rio de Una of Brazil, and Multi-Trex of Nigeria.

Finally, we thank Elsevier Academic Press and its staff for excellent editing and publishing support during the preparation of this volume.

Yuan Zhou
Suresh Chandra Babu

CHAPTER 1

Introduction

Yuan Zhou[1] and Suresh Chandra Babu[2]
[1]Syngenta Foundation for Sustainable Agriculture, Basel, Switzerland
[2]International Food Policy Research Institute (IFPRI), Washington, DC, USA

Contents

The economic growth of nations depends on the growth of their key productive sectors. Most developing countries, where poverty is at the highest levels, rely on the agricultural sector for growth. The productivity of the agricultural sector in turn depends on how farmers use appropriate technologies. Technological change is primarily knowledge driven. Moreover, increasing the productivity of farming systems continues to be the main prerequisite for agricultural transformation. A major constraint to enhancing smallholder productivity is the availability of information and knowledge to address current and emerging challenges at the individual farm level. Thus, agricultural extension systems have played a crucial role in increasing productivity in several developing countries, particularly during the Green Revolution period. Because of recent concerns about global food security resulting from high and volatile food prices, there is now renewed interest in agriculture and support services. Extension figures prominently among those services, along with credit, access to inputs, crop insurance, and links to markets.

The role of extension in enhancing productivity has been well recognized. The major purpose of extension services is to disseminate advice to farmers on a timely basis. Advisory services along with quality inputs are essential tools to enhance productivity. But their optimal use requires knowledge based on research. Knowledge gaps contribute to yield gaps. In addition to technological solutions, farmers also need information on prices and markets, post-harvest management, produce quality requirements, and safety standards. There is increased evidence that large-scale farmers access knowledge through a variety of sources, whereas the "resource-poor" farmers who constitute the majority in rural areas of developing countries continue to have poor access to extension services delivered by governments or other providers

Knowledge Driven Development.
DOI: http://dx.doi.org/10.1016/B978-0-12-802231-3.00001-2

1

to complement their local knowledge (Ferroni and Zhou, 2012). Debate continues as to the extent to which the public sector should provide extension services and when and how privately provided extension system could complement, and in some cases substitute for, the gap left by the public extension system. A key policy question is how to increase the productivity and income of smallholder farmers through an optimal combination of public and private extension approaches, which often coexist in several developing countries.

Extension has a complex history and a mixed record of success among developing countries. Even in countries where public extension used to be effective, at least for a period of time (e.g., in the 1970s and 1980s in China, or during the Green Revolution in India), it is no longer the case, for a variety of reasons. In countries that experienced structural adjustments in the 1980s and 1990s, public extension systems faced drastic cuts in public financing, and this led to a rapid deterioration of extension services. Partly due to this neglect, the public provision of extension has in general fallen short of expectations. Links between research, extension, and farmers are seen to be inadequate. Public extension systems, particularly those recently being decentralized, often face budget problems in terms of both volume and distribution. There are issues with capacity, motivation, competence, performance, and accountability of extension institutions and their agents, which can lead to unsatisfying results. Most public extension systems are unable to reach a majority of farmers with a variety of needs.

It is not surprising that other providers are entering the territory of extension. A pluralistic approach to extension service provision is a widely recognized response to meeting the rising complexity of farmers' information needs. The joint involvement of the private sector, nongovernmental organizations (NGOs), and farmer-based organizations in the provision of extension services is growing rapidly. For example, according to a national survey in India in 2005, most farmers got their information from other progressive farmers and input dealers (17% and 13%, respectively) as well as media, while public extension reached only 6.4% of the farmers (NSSO, 2005).

Agricultural extension provided through the private sector is advancing rapidly in a number of developing countries. These efforts involve seed and input companies, distributors and dealers, service providers, food processors and retailers, and mobile operations and their business partners. Firms are motivated to enter extension to increase sales or revenues from contract farming. Contract farming is an increasingly important vehicle for "embedded services," information tied to input sales or marketed produce (Feder et al., 2011). However, the specific role and function of the private sector and whether and how they work with other partners such as the public sector, farmer groups or associations, or NGOs are not well understood and documented. We often know even less about the incentives that each of these actors faces when participating in the joint provision of the services, such as proper funding, fees for services or cost recovery, and profits.

RATIONALE AND AIMS OF THE BOOK

A book on private sector provision of extension is of interest to development practitioners for several reasons. First, it is useful for the policymakers to learn the differences between private and public extension systems in terms of effectiveness and cost-efficiencies. Second, extension researchers have suggested that there is a need for adequate regulatory mechanisms to protect smallholder farmers who are serviced by private extension agents, as the private companies have divergent motivations (Birner et al., 2009). What regulatory systems and institutional arrangements should be developed to simultaneously support the development of private extension and ensure that the associated farmers are treated fairly? It is important to have options for institutional arrangements that respect both the farmer and the contracting companies.

Third, it is important to learn what institutional innovations have emerged at the farm level to improve knowledge sharing and how such innovations can be used to improve extension in other commodities and other production systems (cross-production system learning for extension). Also, understanding the success of private businesses in reaching farmers could aid the public sector (as well as competing companies) to improve their own efficiency and yield a stronger system overall.

Fourth, understanding the motivation of private companies to engage in knowledge sharing and extension has important implications for the public extension system. Understanding the incentives of the private sector can help determine the extent to which farmers will be served by the companies and hence indicate where and how the public sector must provide complementary services.

Fifth, it is beneficial to understand how private companies integrate outsourcing and contracting decisions with knowledge-sharing functions. If the public extension system functions well, the private sector stands to benefit. In the absence of such well-functioning public extension, private companies must fill the gap and invest in knowledge development and sharing in order to ensure they receive quality inputs for their own enterprises. Conversely, there may be a role for entrepreneurial knowledge workers to sell their services to farmers who have contracts with companies to ensure they are able to supply the desired product (Dewhurst et al., 2013).

Finally, a set of case studies on private sector extension may yield insight into inclusive innovations that involve farmers, local credit institutions, public research organizations, and other private input suppliers, offering the potential to improve the productivity of the system as a whole and benefit the farming community. Inclusive innovation may help smallholder farmers enhance their productivity more effectively than the traditional linear model of research and extension. However, they need to be supported by strategic policy interventions such that the two approaches work in complementarity. Recent studies indicate that policymakers could play a significant role in shaping and supporting inclusive innovation systems (Foster and Heeks, 2013a,b). Yet little attention has been paid to nurturing such innovations through programs or policies.

The specific objectives of this book are as follows:

- to better understand different models of extension delivery and financing by private companies across the agricultural value chain;
- to provide an assessment of the factors leading to success or failure of various approaches; and
- to draw lessons and recommendations for future endeavors.

The findings will also help understand how the public sector can better work with or support private companies to achieve development and extension goals.

KEY RESEARCH QUESTIONS

Embedded extension services in input supply and contract farming by the private sector could work well from small to large farmers in well-endowed areas. But such services are often targeted at high-value commodities, and thus may not apply to other crops, cereals in particular. How wholesale buyers of cereal get involved in specific aspects of the value chain—for example, processing and quality control of cereal marketing and provision of extension services—remains unclear.

With some food grains, where there is limited marketed surplus and the scope for contract farming is small, the private sector may not have any incentive to take a lead role—or any role—in extension, whereas in integrated value chains such as horticultural products for export, the private sector's incentive to collaborate with farmers in a tightly knit contract farming system, and therefore to provide extension, is likely to be very high. Identifying the factors that do not allow private-sector approaches to be scaled up, particularly for cereals, is important. In some cases, even for "buyback schemes," unless the farmer is charged, the cost of extension may not be fully recoverable.

In Africa, where large farms coexist with smallholders, there have been innovative value chain approaches that link the two types of farms, often in the form of "out-grower" schemes, where larger farms or their corporations serve to help smallholders with input provision and processing while also acting as a source of extension services. This is yet another type of private extension that needs further exploration and analysis.

Enhancing the role of private extension to fill the gaps in public extension systems requires full understanding of what works and how in private extension systems. To our knowledge, a number of research questions related to private extension remain to be addressed:

1. What has been the role, behavior, and growth pattern of the private sector in extension?
2. What are the objectives of private extension activities, and what are the incentives for undertaking private extension (to sell a product, to assist associated farmers, etc.)?
3. What is the content of private extension activities (technical support, assistance with input supply or marketing, etc.), and what can be said about the quality of the extension advice?

4. What are the specific methods or approaches used by the different types of private providers of extension, by region or by types of farmers? What level of expenditures is devoted to extension?

5. How successful are private-sector extension efforts in terms of yield increases, quality improvement, or other relevant measures?

6. What is the partnership model (if applicable) and what are the functions of each partner (private, public, NGOs, etc.); what are the incentives they face, and the results and impact achieved?

7. What are the systems of knowledge flow? For example, where do the private extension agents get their knowledge? Are they current? How do they update their knowledge?

8. What role does research play in enhancing the role of private extension? Is this a constraint for the development of the private sector?

In specific settings, there are several other interesting aspects worth investigating, such as how governments, NGOs, and farmer-based organizations respond to the increasing role of private companies; how best to place and facilitate the function of the private sector in a pluralistic extension system; how to co-innovate and generate synergies across different extension actors; and how to align objectives, performance criteria, results, and outcomes from various sectors. These questions will inevitably be touched upon when discussing specific cases.

RESEARCH METHODOLOGY

Ten case studies on private extension are presented in this volume. Some of them fall under the category of input suppliers; others belong to the group of output buyers or aggregators. In some cases, private companies act on both ends, providing inputs and buying back the produce. The cases include EID Parry's sugarcane project in south India; Jain Irrigation's onion program in India; Kenya Horticultural Exporters's French bean scheme in Kenya; Multi-Trex's cocoa farming in Nigeria; Syngenta's Frijol Nica program in Nicaragua; Rio de Una's vegetable operations in Brazil; Marico oilseed production in India; Sarveshwar Organic Foods in Jammu, India; PRAN Vegetable Production in Bangladesh; and AGPP Joint Stock Company in Vietnam. The case studies aim at understanding the effectiveness, efficiency, sustainability, and impact of the private sector–led extension approaches.

The research is conducted using various data collection instruments, including desktop research, literature review, and most importantly field interviews with companies, farmer beneficiaries, and other important stakeholders. Questionnaires for each category have been designed and administrated to solicit the necessary information and data on (1) extension delivery and financing approaches, (2) quality of services, (3) the partnership model, (4) results and impact generated in both qualitative and quantitative measures, and (5) success factors and limitations.

OVERVIEW OF CONTENT

The content of this book is built around actual case studies that were written specifically for the study of the role and capacity of private companies in agricultural extension. Description of specific extension models and approaches are often teased out of a complex situation that exhibits a range of agricultural, regulatory, and socioeconomic variables. Illustrative cases typically focus on a particular agricultural value chain and elaborate the special features of the associated private extension system. In addition to this introductory chapter, the next two chapters provide a comprehensive review of the literature on private extension as well as an overview of the historical development of pluralistic extension systems. Chapter 2 also presents a conceptual framework for analyzing the role and contribution of private extension systems.

Chapters 4 through 13 feature the individual cases of private extension. The chapters follow a similar structure in content formulation. Each begins with a section describing the background and agricultural context of the case, followed by a description of the specific crop value chain. Based on understanding of this context, extension models and methods used by private companies receive deeper analysis and definition. This leads to a discussion of the private extension with respect to its relevance, efficiency, effectiveness, equity, sustainability, and impact. Following that, comparison with public extension and the uniqueness of the extension model, as well as lessons for its replication and scaling up are elaborated.

The final chapter summarizes the major results from the 10 cases presented. It looks at the trends, commonalities, and differences among various extension approaches and teases out the general lessons for success or failure. Then it presents a set of lessons around value creation, integrated services, market links, inclusive innovation, and capacity development.

REFERENCES

Birner, R., Davis, K., Pender, J., Nkonya, E., Anandajayasekeram, P., Ekboir, J., et al., 2009. From best practice to best fit: a framework for designing and analyzing pluralistic agricultural advisory services worldwide. J. Agric. Educ. Ext. 15 (4), 341–355.

Dewhurst, M., Hancock, B., Ellsworth, D., 2013. Redesigning Knowledge Work. Harvard Business Review, January–February 2013.

Feder, G., Birner, R., Anderson, J.R., 2011. The private sector's role in agricultural extension systems: potential and limitations. J. Agribusiness Dev. Emerg. Econ. 1 (1), 31–54.

Foster, C., Heeks, R., 2013a. Conceptualizing inclusive innovation: modifying systems of innovation frameworks to understand diffusion of new technology to low-income consumers. Eur. J. Dev. Res. advance online publication, 4.

Foster, C., Heeks, R., 2013b. Analyzing policy for inclusive innovation: the mobile sector and base-of-the-pyramid markets in Kenya, innovation and development. Eur. J. Dev. Res. 3 (1), 103–119.

Ferroni, M., Zhou, Y., 2012. Achievements and challenges in agricultural extension in India. Global J. Emerg. Mark. Econ. 4 (3), 319–346.

NSSO, 2005. Situation assessment survey of farmers: access to modern technology for farming. Report No. 499(59/33/2), National Sample Survey Organization, Ministry of Statistics & Programme Implementation, Government of India.

CHAPTER 2

The Current Status and Role of Private Extension: A Literature Review and a Conceptual Framework

Suresh Chandra Babu, Natarajan Ramesh and Caitlin Shaw
International Food Policy Research Institute (IFPRI), Washington, DC, USA

Contents

INTRODUCTION

The role of the private sector in providing agricultural extension and advisory services has been well recognized as a means of increasing productivity in the effort to meet food security goals. Agricultural extension providers provide not only technology, but also the information and skills necessary to increase production in a sustainable manner. Traditionally, extension has implied training and dissemination of information surrounding specific production technologies. More recently, it has expanded to include helping farmers to form groups, deal with the marketing of products, and partnering

Knowledge Driven Development.
DOI: http://dx.doi.org/10.1016/B978-0-12-802231-3.00002-4

with relevant service providers such as rural credit institutions. However, in today's rapidly changing environment, it is important to understand the role of pluralistic extension systems in meeting the diverse technology and information needs of farmers. The pluralistic form of extension, combining a variety of service and technical providers, offers a unique opportunity to address the challenges facing smallholders as the agricultural sector transitions to a more market-oriented system.

Recently, there has been growing interest from large international actors in the role of extension in meeting targets such as the United Nations Millennium Development Goals. In addition, the creation of targeted advisory groups such as the Global Forum for Rural Advisory Services has put agricultural extension at the forefront of policy discussions.

In this chapter we review the current status and the role of private extension programs in developing countries. We specifically address the following questions to better understand the role of private extension provision:

- What roles do the private extension systems play in providing advisory services and how do they fill the gaps left by public extension systems?
- What factors determine the entry, establishment, and sustainability of the private extension systems?
- What issues, constraints, and challenges do private extension providers face in developing countries?

As will be seen in the later chapters of this volume, the demand patterns of farmers are changing in developing countries due to the transformation of the agricultural sector. As the income of smallholder farmers rises, farmers are shifting away from cereals to higher-value crops. Production systems have become more specialized, requiring more context- and commodity-specific extension services for farmers, requiring a move away from the top-down, linear approach of the training and visit (T&V) system (Anderson and Feder, 2004). Public extension systems can no longer adequately provide extension because of the limitations of a bureaucratic, "one-size-fits-all" approach and because of the fiscal unsustainability of a government trying to meet all the various, specialized needs of farmers.

As a solution, many countries have privatized some of their extension services—for example, providing vouchers as in Costa Rica and China, or contracting out extension as in Chile. In addition, the number of farmers' associations has increased significantly, enabling smallholders to band together in order to fund private, fee-for-service extension. Furthermore, farmers are also able to obtain extension services (among other services) through contract farming or extension embedded in their transactions with input dealers. However, further research is needed to analyze the true impact of the extension component of contract farming. Yet there is comparatively little evidence on how private extension operates, who benefits, what costs are incurred, and how to scale up these operations. In this review we analyze the literature around how private extension has been emerging in different contexts and the current status, challenges, and constraints on expansion.

The literature review is organized as follows. In the next section we review the emerging trends in a chronological fashion. This is followed by literature on the privatization of public extension. Then we review private extension models that have been modified for the specialized information needs of farmers. We then look at the issues around analyzing private extension from a policy perspective.

The literature review is followed by development of a conceptual framework that guides the case studies presented later in the book, in terms of identifying the key elements in assessing the role and impact of private extension services. The chapter ends with some concluding remarks. Following the end of the chapter, Table 2.3 provides a detailed list of publications on private extension and their areas of focus.

TRENDS IN THE DEVELOPMENT OF PRIVATE EXTENSION

After an extensive literature review of journal articles, discussion papers, and other articles from 1994 to 2014, several trends have emerged. Current overarching trends in agricultural extension are broadly focused on decentralization, outsourcing, and privatization. Among the most prominent trends are the following:

1. Pluralism and decentralization
2. Gradual transitioning from public funding to subsidies to direct payment
3. Contract farming
4. Embedded extension from input dealers or mechanical suppliers
5. Public–private partnerships

Another important trend that emerged was the gradual change in focus from using welfare economic ideas to create a framework for analyzing private extension to incorporating a more diverse set of ideas to build an analytical framework, such as sociological network theory, institutional economics, and systems theory (Labarthe, 2009; Faure et al., 2012).

Pluralistic agricultural extension systems

Pluralistic agricultural extension and advisory services refers to the growing variety of services providers operating in this space. This extension structure acknowledges the need to address agricultural and rural development challenges with varied approaches (Heemskerk and Davis, 2012). This encompasses the outsourcing and privatization trends mentioned above to include public–private partnerships, farmer-based and nongovernmental organizations (NGOs), as well as private input suppliers. The benefits of a pluralistic extension system include the ability to overcome common constraints such as funding, personnel, and capacity as well as allowing the flexibility to meet the specific needs of different subgroups of farmers or regions. This allows the system to better meet the information, skill, and technology needs of the farmers. It is recommended that the public extension system act to coordinate and manage the stakeholders toward the common goal of addressing the needs of farmers in order to capitalize on the specialized roles of each organization.

Privatization of public extension literature

One of the earliest frameworks for the privatization of extension was based on a concept from welfare economics: the rivalry and excludability of goods (Umali and Schwartz, 1994; Umali-Deininger, 1997). The authors combined these characteristics to sort the types of agricultural information into public (non rivalrous and non excludable), private (rivalrous and excludable), common-pool (non excludable and rivalrous), and toll goods (excludable and non rivalrous) (Table 2.1). This characterization enabled the authors to answer their initial question about who should fund agricultural extension. They concluded that farmers will only pay for private goods and information that can be characterized as a toll good (basically only non excludable information).

Decentralization emerged as a popular theme in the literature as both a cause and effect of privatization. Many countries, such as Chile, privatized their extension services as part of the government's more general fiscal decentralization. The Chilean decentralization of extension also had the effect of "improving client orientation and ownership" (Umali-Deininger, 1997). Decentralization has also helped extension services to be more responsive to the needs of farmers (Birner et al., 2009). Rivera and

Table 2.1 Economic classification of agricultural information/technologies transferred via extension

	Excludability	
	Public goods	**Private goods**
Low	• Long-term general agricultural information • Large-scale information dissemination (e.g., market prices)	• Pure (general) agricultural information (e.g., cultural and production practices) • Specialized agricultural information (e.g., farm management, marketing, processing)
Rivalry High	• Improved agricultural technologies (e.g., fertilizer, improved seeds, irrigation)	• Modern technologies (machinery, chemicals, hybrid seeds, etc.)

Source: Adapted from Umali and Schwartz (1994).

Alex (2004) also emphasize the importance of decentralization but point out that without effective monitoring and coordination the effectiveness of these services is blunted. More radical forms of decentralization suggest the complete withdrawal of government, but this has proven to be ineffective (Rivera and Sulaiman, 2009). On the other hand, local governmental regulation is a moderate possibility with the merits of decentralization and public sector monitoring/regulation (Birner and Anderson, 2007). The literature also suggests that decentralization be used as just one tool in a "menu of options" in the process of extension privatization. Decentralization may also lead to problems of political interest capture and create incentives to burden extension agents with non extension tasks, limiting their effectiveness in providing advisory services (Birner et al., 2009). Decentralization was also one of the four themes agreed upon by a 2002 extension-related workshop convened by the World Bank, USAID, and the Neuchatel Group (Rivera, 2006). However, decentralization can lead to needless redundancy because of a lack of coordination (Rivera et al., 2002). Faure et al. (2012) point out that decentralization improves advisory service systems and provides sounder advice than government advisors. A major negative side effect of decentralization is that it reduces economies of scale, making extension less effective for smallholders (Benson and Jafry, 2013). This study further cites the need for diversified service providers. Table 2.1 illustrates the roles of both public and private actors in the provision of agricultural extension. Umali and Schwartz observed that "A central objective in a private fee for service extension system is in getting the right message to the right individual or group through the creation of a demand driven extension service system that is cost-effective, efficient and of high quality" (1994). The authors categorize agricultural information in two ways: pure agricultural information, and agricultural information which is attached to an innovative technology. First, pure agricultural information can be used without requiring the user to have advanced technology. This includes improved production techniques such as those used in land preparation, planting and harvesting, or the optimal level of input use. Information embodied within agricultural technologies includes marketing and processing technologies as well as modern agricultural inputs such as improved varieties and chemicals.

The consensus of the literature is that pure private fee-for-service extension (without any public sector involvement) is both impractical and undesirable. The exception to this is embedded extension. The literature also generally agrees that a pluralistic system is best. Where the literature diverges is on what roles the public and private sectors should play in order to meet the needs of all farmers, especially smallholders. Some authors argue that the public sector should provide public extension-related goods while the private sector should provide private extension-related goods (Umali-Deininger, 1997). This division implies that private providers will primarily serve commercial farmers with large landholdings who farm high-value crops, while the government will serve marginalized farmers (Muyanga and Jayne, 2008).

An oft-proposed alternative to having the government serve smallholders and other marginalized farmers is to facilitate farmer cooperatives to take advantage of economies of scale and afford private extension (Umali and Schwartz, 1994). Also, a public good can become a private good if the people who would be affected by the externalities help pay for it through collective action (Feder et al., 2011). In addition, a stochastic analysis in Crete shows that public and private extension are more efficient together than separately (Dinar, 1996). Further, when one sector finances the extension system (usually the public sector) and the other delivers the services (usually the private sector) the system as a whole has been seen to be more effective (Birner and Anderson, 2007). However, this is not to say that the two systems should work in a mutually exclusive manner in two silos. For example, even when the private sector provides and delivers extension services, the public sector often needs to play a regulatory role to ensure quality control (Umali-Deininger, 1997). In addition, when several positive externalities are at stake (such as potential for improving soil quality and teaching farmers environmentally sustainable farming practices) the public sector should be involved in order to spread this knowledge more widely (Rivera, 2006). Picciotto and Anderson attribute the rise of the pluralistic paradigm to the growing influence of New Institutional Economics, especially Umali-Deininger's work; the strained capacities of the public sector; the growing need to provide specialized extension for specialized needs; and the development of new information and communication technologies (ICTs) (1997). The cases of Nicaragua and Uganda show that privatization of extension is less successful if public extension still operates in parallel (Feder et al., 2011). This is partly due to the public and private sectors competing to serve the same function, as opposed to serving complementary functions. Privatization is more successful if there is a plurality of providers *within* the public sector as well (Chipeta, 2006). It is also important that the government provide a solid infrastructure so that private providers can reach marginal areas, thereby increasing the effectiveness and impact of their efforts (Muyanga and Jayne, 2008). Although pluralism is beneficial, if the public sector views the new private providers as rivals or there is confusion over the respective roles of each sector, its benefits are greatly reduced (Benson and Jafry, 2013). Therefore, it is important that these roles be clearly defined and adhered to in order to promote coordination.

Another recurring theme in the literature is that of gradualization. In the literature, gradualization usually refers to the deliberate withdrawal of the public sector and often additionally refers to a slow entrance by private providers. Gradualization often manifests itself in the form of gradually decreasing subsidies (Hellin, 2012). Rivera and Alex's analysis of several case studies led them to conclude that privatization is ineffective if it is not gradual (2004). Furthermore, Adejo et al. argue that in the case of Nigeria, the public sector should withdraw gradually from commercial farming areas but not withdraw *completely* (e.g., by contracting instead of delivering extension) to marginalized areas (2012). Another form of gradualism, that can help to ensure the smoothest possible transition is the use of pilot programs (Kidd et al., 2000). In Tasmania, Australia, there was a gradual transition from a commercial ministry to a state-owned enterprise with the state as the initial sole shareholder. On the other hand, an abrupt change in New Zealand resulted in a reduction in professional staff and clientele of more than 50% (Bloome, 1993). Also, Saravanan stated that the public sector (specifically the Indian government) should withdraw gradually either area-wise (areas with favorable environments are privatized first) or commodity-wise (commodities with high profit margins are privatized first) (2001). Finally, exclusion of the gradual coalition in Uganda led to suboptimal outcomes and social capture (Rwamigisa et al., 2013).

The importance of group-building in the privatization of extension is stressed in the literature. Facilitating the formation of farmer or producer groups helps articulate demand, making extension more demand driven (Birner and Anderson, 2007). Group formation also helps link smallholders to markets (Ferris et al., 2014). Some authors see group-building as the responsibility of the public sector because they perceive farmers' cooperatives as a public good (Kidd et al., 2000; Chipeta, 2006). Collective action by farmers also takes advantage of economies of scale, making private extension more affordable for smallholding farmers to buy and more justifiable for private extension providers to provide (Benson and Jafry, 2013). Farmers' groups are often difficult to form for cultural reasons (Hellin, 2012). Marginalized people and women should be incorporated into groups through a form of affirmative action (Okoboi et al., 2013). The importance of group-building and formation of cooperatives is widely agreed upon (Adejo et al., 2012).

Many authors argue that the privatization of extension should be demand driven, or alternatively one of the goals of privatization should be to make extension more demand driven. To make extension more demand driven, a market mechanism is necessary. This means that either extension should be purely private or there should be a public–private partnership that creates a proxy market mechanism through means such as contracting or giving out vouchers (Birner and Anderson, 2007). One of the problems with fostering demand-driven extension is that it is often difficult for farmers to articulate demand. As mentioned above, group formation is important to support the articulation of demand, especially in smallholders. On the other hand, pure demand-driven extension often fails to address the objectives of the state, such as promoting environmentally sustainable practices (Knickel et al., 2009). Faure et al. argue that extension cannot be reduced to being demand or supply driven but must be based on interactions between farmers and advisors (2011).

Another trend in the privatization literature is a focus on market-driven extension, or extension that focuses on improving the linkages between actors in an agricultural innovation system (AIS). Rivera and Sulaiman point out that extension has evolved from production-focused to group organization–focused to market linkage–focused (2009). Muyanga and Jayne's study emphasizes that marketing strategy advice was one of the least often adopted, but most important, good practices on their list (2008). One of the most important functions of extension in the eyes of a smallholding farmer is linkage to markets, because this enables the farmer to move from subsistence farming to commercial farming (Ferris et al., 2014). The next section will focus on the models of extension privatization implemented, and the challenges and opportunities surrounding each.

Private extension literature

The earliest literature on private extension was a short series of papers by William M. Rivera. His analysis of privatization in France, the United Kingdom, the Netherlands, and New Zealand led him to classify private extension into three models: public financing by taxpayers for material relevant to the general public, with the rest of the funding coming from direct payment from farmers, as seen in France; direct charging of users without privatization of provision, as seen in the United Kingdom; and an equal distribution of labor between the public and private sectors, with responsibility for coordination belonging to the public sector, as seen in the Netherlands (Rivera, 1992).

Rivera also concluded that privatization has negative effects on smallholding farmers because providing extension service to them is unprofitable for private providers, especially given their geographic disparity and small farm size (Rivera, 1992, 1993). On the other hand, although Rivera and many other authors say that private extension is not an adequate solution to help alleviate the poverty of smallholders, they do not reject the notion of private extension completely (Faure et al., 2012).

Contract farming is one of the key types of private extension found in the literature review. Contract farming both guarantees a market for the farmer and helps the provider ensure the quality of his produce, making it beneficial to both parties and one of the most effective forms of private extension (Ferris et al., 2014). In India, contract farming has proved particularly potent because it unites the front and back ends of extension. Although extension is not always part of contract farming, Minot makes a strong case that it should be, because of the mutually reinforcing nature of the two activities in his analysis of several case studies conducted by Cornell (2007).

Public–private partnerships are a type of extension that came about as a result of a focus on pluralism in privatization. For example, in Nicaragua the government provided vouchers to farmers to pay for private provision of extension from one of the providers (Umali-Deininger, 1997). Costa Rica also implemented a voucher system (Dinar, 1996). Another form of public–private partnership is subsidy. In the Netherlands, private extension was 50% subsidized by the government (Rivera, 1993). India's ATMA reform also encouraged public–private partnerships (Sulaiman, 2012). In Chile, public financing and private provision in the form of subcontracting was used to provide extension to smallholders.

The public sector can also serve an important purpose as a certifier to ensure that the quality of private extension providers is adequate and that farmers are not misled. For example, India's DAESI (Diploma in Agricultural Extension Services for Input Dealers in India) is helpful for farmers trying to choose an input-dealing extension provider. Ghana has an equivalent program called GAAD (Ghana Agro-Dealer Development project) (Salifu et al., 2010). This certifying function of the public sector satisfies the prerogative (Umali-Deininger, 1997) for the public sector to play a role in regulating the extension market, while letting the market mechanism inherent in private extension take effect.

ICTs are playing an increasingly important role in the provision of private extension. Taking advantage of new ICTs is essential to providing private extension (and making it attractive for providers to do so) in marginalized and hard to reach areas, especially when infrastructure and/or personnel are lacking (Howard et al., 2012). The literature seems to be in general agreement that ICT is a useful tool, but definitely not a silver bullet that will solve all the problems associated with in-person contacts (Howard et al., 2011; Richardson, 2006). The Indian Tobacco Company's e-Choupals (electronic kiosks that provide extension run by a local farmer) and Digital Green (which combines a website and human advisors) are excellent examples of the potential of ICT. They also demonstrate that ICT is a tool and not a solution in and of itself (Sulaiman, 2012). Mobile extension also shows a lot of promise but doesn't help the extremely poor (Ferroni and Zhou, 2012).

Embedded extension provided by input dealers and other input actors or output aggregators or other output actors serves a function somewhat similar in implementation to contract farming but has proven to be a useful and widely used model for private extension. Tata Kisan Sansar's extension program was rated as medium in effectiveness by 54% of farmers, whereas the other 48% rated it high in effectiveness. Transmar Ecuador's program of embedded extension helped farmers take advantage of economies of scale through vertical integration, get rid of the middleman so they can have a greater share of the value chain in cacao farming, and meet the quality demands of their consumers (Blare and Useche, 2014). PepsiCo's high-quality seed program in India and the Indian Tobacco Company's e-Choupal program are also examples of embedded extension that have been successful (Sulaiman, 2012).

Although not strictly extension providers, innovation intermediaries, network brokers, and other brokers and facilitators serve an important role in private extension (*note:* extension workers can often double as facilitators or incorporate facilitation into their work). Facilitators support the co-creation of knowledge by farmers and researchers, as opposed to the linear transfer of technology. They are an integral part of an AIS (Faure et al., 2012). The work of Klerkx and Leeuwis has been especially critical in arguing for the importance of so-called "Innovation Intermediaries" who help articulate demand, serve as cultural and cognitive bridges, function as liaisons between

large providers and small farmers, and address the asymmetrical distribution of information between providers and farmers (Klerkx and Leeuwis, 2008). The *Kamayoq* and
despacho models from Peru and Mexico, respectively, demonstrate how network brokers
can be used to make it feasible for marginalized areas/groups to benefit from fee-for-
service extension. NGOs often make good network brokers as well (Hellin, 2012).

Based on the above literature review, we next present a conceptual framework for
analyzing the role and impact of private extension services in agricultural development.

CONCEPTUAL FRAMEWORK FOR THE ANALYSIS OF PRIVATE EXTENSION

The principles of private provision of public services (e.g., health and education) to
communities similarly apply to the provision of agricultural extension and advisory
services. Issues related to the quantity and quality of services and their delivery, who
pays for the services, and how the service provider is held accountable are common
to any social service provision. Yet the provision of extension services through private-
sector entities has specific issues that must be considered in order to analyze their functioning. Further, understanding these principles in the context of a broad conceptual
framework will help in the development of specific hypotheses that can be tested in
the analysis of the private extension provision.

This volume is aimed at gathering a wide range of experience around the functioning of private extension systems through specific case studies of providers. Private
extension often exists in one form or another along with the public extension system, either through formal advice given during the sale of seeds or agrochemicals or
through some specialized approach coupled with crop or livestock production. The
case studies in this volume go beyond the incidental role private actors play and the comprehensive approach the private sector takes in offering full-scale extension services
through input supply, output aggregation, or both.

The purpose of the case studies presented in this volume is to collectively inform
policymakers in developing countries and development partners willing to invest
in the development of the private extension system. To best inform this audience,
we focus on the actions of private entities that are engaged in advisory services and
how they interact with public-sector actors, NGOs, and farmers, whether directly or
through farmer organizations. To further supplement the case studies we have drawn
heavily on the literature which looked at the historical development of the concept of
private extension and distinguished it from attempts to privatize the public extension
systems through various forms of incentives.

Any attempt to understand the role and functioning of private-sector provision of
extension must consider several contextual factors at the country, regional, and local
levels. Several policy-related questions need to be answered: How can one promote

private extension without undue exploitation of farmers by the private sector? How do the policymakers and program managers coordinate the entry of the private sector for specific crops in the presence of the public-sector extension provided in several developing countries? How does one manage the coexistence of public and private extension for the same production systems under similar agroecological conditions? Under what conditions might there be effective development of the public–private collaboration? How might several actors and players come together, crossing their boundaries—such as irrigation companies, agrochemical dealers, credit institutions, and the private extension system—to promote multi actor support and advisory services to farmers?

The public–private extension continuum—a framework

Understanding the current status of the extension pathways in an agroecological region or an administrative region requires mapping out the players and actors in the area and studying the intensity of their involvement with the various types of farmers as well as their interaction with those farmers. At one extreme, one might see that all the extension programs are organized and implemented by the public extension system with full funding from the government. At the other extreme, one could imagine a system where full extension services are offered by the private entities based solely on demand for their services. However, most of the situations one will study will have a combination of some level of not only public and private extension operators providing services at differing degrees of intensity, but also other actors such as FBOs, NGOs, and CSOs playing their roles in their specific areas of interest to help the farmers. We borrow these ideas from Timmer's chapter on agricultural transformation contained in the *Handbook for Development Economics*, where the author highlights the need for an effective extension system to support sustainable agricultural transformation (1988). Timmer cites the central role of agricultural extension and advisory services as an "essential ingredient in speeding up the adoption of new agricultural technology" as a means of increasing productivity (1988). A stylized depiction of such combinations during various stages of development is given in Figure 2.1. It is important to note the increasing role of private extension provision as the agricultural sector becomes further transformed. This is due to the increasing diversity of farmers' information and technology needs as the sector becomes more commercialized. It is also important to note the trend of pluralistic extension systems, which is especially important at the medium level of agricultural transformation.

Countries and regions may have different combinations of these extension approaches in actual practice. Understanding the factors that affect the performance of these providers in combination with other related players and actors would help to identify the opportunities to improve systems that are weak and sustain those that are delivering their services effectively. This can further help in the design of context-specific and appropriate extension policies and related incentives that would ultimately

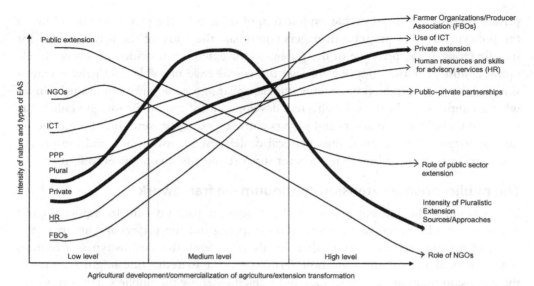

Figure 2.1 Stylized trends in the nature of EAS and extension transformation. *Source: Authors' compilation.*

lead to a well-functioning extension system that is able to support farmers in a sustainable way.

What factors motivate entry, effectiveness, and sustainability of the private sector in extension provision?

Increasing the productivity of farming systems continues to be the main driver of agricultural transformation. The foremost constraint on enhancing smallholder productivity is the availability of information that addresses current and emerging challenges at the farm level. The role of extension and advisory services in enhancing productivity has been well recognized. Davis et al. found that agricultural extension and advisory services are an opportunity to strengthen households by "increasing their access to tangible and intangible resources, such as inputs and knowledge" (2014). The major purpose of extension services is to disseminate knowledge to farmers on a timely basis. Advisory services along with quality inputs are essential productivity-enhancing tools. But their optimal use requires knowledge based on research and innovation. Knowledge gaps contribute to yield gaps. In addition to technological solutions, farmers also need information on prices and markets, post-harvest management, produce quality requirements, and safety standards. There is increasing evidence that large-scale farmers access knowledge through various sources while the smallholders, who form the majority in rural areas in developing countries, continue to have poor access to extension services (Babu et al., 2013). Debate continues, however, on the extent to which the public sector should provide extension services and when and how privately provided extension systems could complement and in cases substitute for the gap left by the public extension system.

There are many reasons behind the need to reform the extension system in many developing countries. First, government funding for public provision of technical and advisory services in the agricultural sector is limited. This calls for doing more with less. Further, public extension systems, particularly those in the process of being decentralized, often face budget problems in terms of both volume and distribution. Second, the quality and coverage of these services is limited and needs to be improved in order to have the intended impact. There are challenges with the capacity, motivation, competence, performance, and accountability of extension institutions and their agents, which can lead to unsatisfying results. Most public extension systems are unable to reach a majority of farmers, who have varying needs. In this context, it becomes imperative to undertake an analysis of how various systems of extension provision are functioning in a country, agroecological region, or local community in order to guide policymaking to improve available extension and advisory services. In what follows, we provide an operational framework for assessing, analyzing, and designing specific extension reforms.

Operational framework for assessing, analyzing, and acting on extension reforms

Figure 2.2 illustrates the operational framework for effectively assessing, analyzing, and acting on extension reforms. This figure highlights the need for assessment and analysis in order to effectively inform policymaking, so that it in turn can more effectively allocate resources to meet the information needs of farmers.

Assessment of the current state of extension

Before any improvements can be made to the private extension system, an assessment of the current state of extension in the country must be conducted. The specific role and function of the public sector and the third sector, and whether or how they work with other partners such as farmers' groups or associations or NGOs, are not well understood or documented. To enhance this understanding, it is necessary to map the roles of various key actors and stakeholders in various extension systems and advisory services currently operating. It is also critical to understand the linkages and relationships between these actors and their stakeholders in order to study the strengths, weaknesses, gaps, and potential in the system.

Several methods may be used to assess the status of the extension system. The quantitative approach would include taking stock of various forms of investments in reaching the farmers through different extension systems and their approaches, including

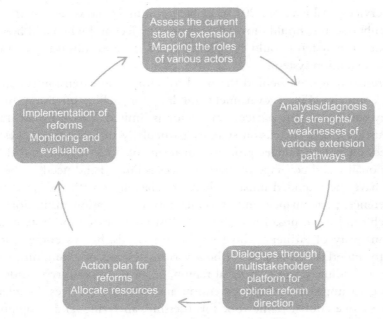

Figure 2.2 Operational framework for assessing, analyzing, and acting on extension reforms. *Source: Authors' compilation.*

human, physical, and financial capacities. In addition, qualitative techniques such as case studies, farmer focus groups, and mapping approaches can be used to describe the systems, along with the potential for collaboration, cooperation, and networking opportunities to deliver cost-effective services to farmers.

Analysis

An in-depth analysis of extension provision is of interest to the development community for several reasons. First, it is useful for development practitioners to understand which farmers can benefit from the system and how. Second, it is useful for policymakers to learn the differences between various systems, such as the public and private extension systems, in terms of effectiveness and efficiency. Third, it is important to understand the motivation of specific actors to engage in extension and advisory activities. Finally, a study of multiple systems will help us understand how the public sector can better work with or support private companies to achieve development and extension goals. In order to remain relevant and address the current needs of farmers, the pluralistic extension system should evolve as the agricultural sector transforms to become more commercialized. Similarly, in order to increase the potential impact on smallholder production, the private extension system's services should align with the country's agricultural transformation objectives as outlined in national policies.

In the framework described below for analyzing the rural extension system, we consider six variables: relevance, efficiency, effectiveness, equity, sustainability, and impact (following OECD 1991, 2011). The relevance of the private extension system stems from the need to address the technical and advisory service needs of all farmers, particularly smallholders. Efficiency in the extension system refers to the ability of the system to provide the intended benefits at the lowest possible cost, in terms of both money and time. An extension system is considered to be effective if it can reach its targeted beneficiaries and provide them with services at the intended level. An extension system may be considered to increase the level of equity in the delivery of extension services since it may address gaps in technical assistance left by other providers such as the public system. The sustainability of extension programs depends on the sustainability of the entity itself and its ability to maintain its competitive edge and hence its relevance and share in a pluralistic setting. As long as the entity continues to show a positive contribution from its operations and is able to offer an attractive service to its farmers, the extension system may be said to be sustainable.

Table 2.2 illustrates the factors that could affect the impact and effectiveness of an extension system. This helps us identify the strengths and weaknesses of various extension pathways. It is important to identify both internal and external factors that influence the performance and impact of extension systems in order to successfully identify and address gaps.

Dialogue and consultations

Optimal reform of the extension system is only possibly through effective dialogue between all key stakeholders in order to coordinate efforts. This requires a clear and efficient dialogue between players to facilitate communication, collaboration, and knowledge sharing among key actors. An effective multi stakeholder platform might include methods such as key informant interviews and focus group discussions. The collection of accurate information from critical stakeholders is important to understand the roles of players in order to conduct analyses and identify challenges and opportunities.

Action plan for reforms

In order to successfully implement extension system reforms, resources must be allocated efficiently to meet the budget and human capital needs necessary to manage the new system and lead to the intended impact. Additionally, in order to see whether or not the reformed system is benefiting the intended group of farmers, a functional monitoring and evaluation system must be in place. This system will measure and track

Table 2.2 Analysis of factors affecting the successful performance of extension systems

	Relevance	Effectiveness	Efficiency	Sustainability	Impact	Equity
Internal factors	• Crop production (quality and timing) • Timing of operations for cultivation • Timely supply of inputs • Timely purchase of outputs • Market advice • Post-harvest handling	• Leadership/vision • Organized teams • Coordination with farmer groups • Collaboration with input dealers and credit institutions • Partnerships with public extension • Research–extension linkages	• Optimal farmers per extension agent • 24/7 call center • 48 h response • Farmer training centers • Internal coordination with cost recovery	• Sustained supply of outputs • Productivity versus land expansion • Integrated pest management • Controlling water irrigation • Value addition • Market for processed outputs	• Profitability • Labor saving • Effective social capital • Network with other service providers • Achievement-oriented professionals • Competitiveness increase	• Differential engagement of farmers for skill development • Differential pricing of outputs • Help smallholders with specific technology packages • Quality control support • Value addition support
External factors	• Location of processing facility • Public policy for employment • Incentives for problem solving • Recognition and public investment	• Presence of research units • Farmer interest groups • Hire purchase investors • Presence of farmer support	• Pricing of input/output policy • Availability of seedlings/inputs • Mechanization policy • Regulation of pesticide use • Road network transportation	• Rainfall changes • Water levels • Droughts/floods • Labor availability • Soil fertility depletion • Input pricing • Output pricing	• Farmer income/livelihood change • Recognition by the government • Ecosystem benefits • Win–win for the farmer and systems	• Public interventions supporting farmers • Banking/credit institutions • Government subsidies • Procurement support from public/private sectors

Source: Authors' compilation.

the performance of the extension system, and the information generated from the M&E system should create recommendations for improvement.

Given the above framework and the indicators for analyzing the performance of extension programs, we develop a systematic approach to study private extension systems and we apply that approach to the various case studies analyzed in this volume. Several factors help motivate the entry of the private sector into the provision of agricultural extension services. We divide them into two groups: internal and external. Below, we expand on a few of these factors and their relation to the analysis of private extension provision.

External factors

To begin with, the broader sociopolitical environment which allows private entities to directly work with farmers is a major factor in determining a system's success. Three decades ago private extension was banned in several developing countries because the government did not want to the private companies providing extension services to "exploit" the farmers. Over the years, as the effectiveness of the public extension declined due to lack of funding and low emphasis on agriculture as an engine of growth, private-sector extension slowly gained momentum as countries realized that the public sector alone cannot address the information needs of all farmers.

The next facilitating factor is the commercialization of agriculture, in terms of both the market orientation of subsistence production (such as cereals) and commercial production of crops such as cotton, sugarcane, vegetables, and fruits. While commercial production of crops is a natural entry point for the private sector to provide advisory assistance to the farmers, subsistence production requires an entirely different set of information, and thus requires a unique form of extension. In addition, subsistence production does not have the opportunity for large profit margins like those in commercialized farming. Therefore, this is an opportunity to involve a different set of private actors such as NGOs and farmer associations.

Next, natural resource constraints such as the availability of water and rainfall patterns also determine the nature of the information and services required for each agroecological zone and farming system. This speaks to the need for private extension service delivery to be highly context specific in order to address local challenges and needs of farmers, particularly smallholders whose livelihoods are most vulnerable to weather shocks. In addition, the structure and availability of labor markets affects the potential for value chain and agribusiness development. Therefore, these factors should be taken into account by extension providers when determining appropriate diversification or intensification methods that are labor intensive.

Research–extension linkages contribute significantly to the impact of private extension systems. Strong linkages between research and educational institutions can provide support in solving emerging local problems. These institutions can also be a source of capacity building for extension agents and institutions in order to increase the impact of service delivery.

Internal factors

Many internal factors affect the performance of private extension systems by affecting both the system and the farmers it aims to serve. In terms of the system's relevance, the timely supply of inputs is crucial to adequately meet the needs of farmers. If the private system is providing fertilizer, for example, it must be delivered in a timely manner so that farmers may apply it at the appropriate point in the production process. In terms of the effectiveness of private extension services, coordination with local farmers' groups can increase the effectiveness of services through communication of local challenges and constraints. In addition, private training centers for farmers can increase the effectiveness of the extension system by reaching a larger number of farmers with a smaller number of agents.

Assessing the need for private extension

Assessment of the need for the private sector to provide extension in the agroecological zones and for specific crops is necessary to determine the amount of resources to be dedicated to a given area. This needs assessment can help to effectively allocate government resources if coordinated with the private sector, and should be a part of the agricultural sector development plan. Policy and institutional incentives could help the private sector to operate effectively and even substitute for the absence of the public sector. The needs of specific farming systems, vulnerable groups such as smallholder farmers, and women farmers need to be addressed in case the private sector is not able to reach these groups or if it is not profitable to reach them. Such consideration will help in negotiations between the public and private sectors to ensure that the needs of all farmers are considered. This assessment will also reveal the need for the human, institutional, and systems capacity needed to support the development of the private sector.

In order to reach the intended farmers and have the intended impact, we must first understand the role of policies and institutions in promoting the private extension system. First, we need to assess whether the agricultural education system and public extension providers are effectively meeting the information needs of farmers. This assessment should examine the types of trainings offered by these providers. Do the coordinating institutions provide vocational trainings to farmers? Do these providers offer agribusiness development services to help farmers transition out of pure subsistence farming? Do the existing institutions strengthen farmers' groups and cooperatives through effective coordination? Then we must identify any gaps in service in order to assess where private extension services will be most useful. What efforts have already been made to improve access to, quality of, and effectiveness of extension services, and how successful were these efforts? After we have identified these gaps we can then match the remaining needs with appropriate private extension services.

In order for the private extension system to remain relevant, it must be strongly linked with both public and private research systems. These linkages will ensure that private extension providers are up to date with the most recent research developments so that these improved technologies, techniques, and practices can be delivered to their client farmers. In addition, strong linkages between research institutions and extension providers help to ensure that the knowledge of extension agents is updated on a regular basis, which will increase the impact of extension services. The private extension system must also link with local agribusiness entities in order to remain relevant. This includes large companies as well as small and medium enterprises to promote technology and knowledge transfer and increase market access.

CONCLUDING REMARKS

The essential transformation of the agricultural sectors of developing countries must be coupled with evolution of the extension system in order to be successful. In order to meet the diverse needs of farmers to increase production and productivity levels, the most efficient system involves the privatization of extension through a variety of actors, in addition to the public sector. This literature review has introduced the methodology behind privatization and pluralism in agricultural extension along with emerging trends. The extensive references collected through this literature review are included in Table 2.3. This table includes publication details, key concepts, and the conclusions reached by the authors. In addition, we introduced an analytical framework that provides a broad base for studying extension provision in developing countries. Such a framework can be helpful in identifying key factors that are to be studied in understanding the issues, constraints, challenges, and possible solutions for promoting effective, efficient, and sustainable provision of extension services through technological, institutional, and policy innovations.

Table 2.3 Summary of recommended literature on private and pluralistic extension

Article title	Author	Publication information	Issues addressed	Conclusions reached
Public and private agricultural extension: Partners or rivals?	Dina Umali-Deininger (1997)	*World Bank Research Observer,* vol. 12, no. 2	Who should fund and who should deliver extension, the public, private for-profit, or private nonprofit sector?	The goal should be a pluralistic extension system, with private delivery and some financing, and public regulation, standard setting, and some financing in the case of significant externalities and/or a public good.
Public and Private Agricultural Extension: Beyond Traditional Frontiers	Dina L. Umali and Lisa Schwartz (1994)	World Bank Publications, no. WDP236	Efficiency and equity of private sector's ability to provide extension? Will it exclude smallholders and rural poor? How to enhance linkages between public–private–NGO extension?	Private sector will mostly provide toll and private goods for medium–large farmers, public will have to help small except when farmer associations exist. Government must ensure macroeconomic stability, good infrastructure, and good legal system.
Extension Reform for Rural Development, Volume 2: Privatization of Extension Systems	William Rivera and Gary Alex (2004)	World Bank Publications; Agriculture and Rural Development Discussion Paper 9	Case studies of private extension and public–private partnerships as well as lessons learned.	(Authors didn't make critical judgment so conclusion are our own.). *Privatization ineffective if not gradual *Privatization must strike a balance between improving profit for commercial farmers and helping smallholders. *Public–private works best for helping rural poor. *Decentralization should go with privatization. *Privatization of extension generally part of larger liberalization.

Title	Author (year)	Journal	Description	Key points
Extension: Object of reform, engine for innovation	William M. Rivera and V. Rasheed Sulaiman (2009)	*Outlook on Agriculture*, vol. 38, no. 3	Covers history of extension reform and tries to prove that extension should now be reformed to meet the end goal of promoting innovation.	★Monitoring important. ★Privatization brings in corporate clients. ★Extension is usually hierarchically organized. ★Social control is important. ★Farmers must be willing *and able* to pay for extension. ★Extension has evolved from production focused to group organization focused to market linkage focused, as well as from monolithic to pluralistic. ★Public sector realized that it needed to be somewhat involved after trying complete privatization. ★Extension pluralism is desirable. ★Poor farmers must organize into groups to take advantage of private extension.
From "best practice" to "best fit": A framework for designing and analyzing pluralistic agricultural advisory services worldwide	Regina Birner et al. (including Ephraim Nkonya) (2009)	*Journal of Agricultural Education and Extension*, vol. 15, no. 4	Which forms of providing and financing agricultural advisory services work best in which situation?	★Birner's "Best Fit" approach is better than one-size–fits all extension reform. ★Extension reform must change from a "one-size-fits-all" approach to a "best fit" approach that utilizes a "menu of options." ★See framework screenshot on USB. ★Transaction costs concepts of New Institutional Economics should be used instead of Welfare Economics matrix between rivalry and excludability. ★Framework can be used to develop an assessment tool for extension analogous to QSDS for edu/health services.

(Continued)

Table 2.3 (Continued)

Article title	Author	Publication information	Issues addressed	Conclusions reached
Evaluating the impact of agricultural extension on farms' performance in Crete: A non-neutral stochastic frontier approach	Ariel Dinar, Giannis Karagiannis, Vangelis Tzouvelekas (2007)	*Agricultural Economics*, vol. 36, no. 2	Creating a framework to integrate the production- and efficiency-based approaches to impact evaluation in agricultural extension.	★Public and private extension are more efficient together than separately, implying that they serve different purpose and needs of farmers (*note*: more research is needed because the extension was measured only by number of visits). ★Young farmers are inclined toward private extension to close the tech gap, whereas old farmers inclined toward public sector. ★Agrees with recommendations of Umali–Deininger (1997).
How to make agricultural extension demand-driven? The case of India's agricultural extension policy	Regina Birner and Jock R. Anderson (2007)	International Food Policy Research Institute (IFPRI), Discussion Paper 00729	What is each sector's (public, private for-profit, NGO/FBO) role in providing and financing agricultural extension? How to meet the new challenge of helping smallholders access global markets? How can extension address the needs of woman, poor, and disadvantaged? How to keep extension providers accountable to farmers?	★Any solution that addresses market/state/third sector failures comes with its own set of challenges. ★A combination of two different sectors (one for providing and one for financing) works better. ★Market mechanism or a proxy key to ensuring demand driven. ★Linkages/feedback between government, farmers, NGOs, FBOs, and for-profit companies must be strengthened. ★Capacities of extension agents and farmers (to organize in groups) must be improved. ★Farmer-to-agent ratio is >1,000:1, →FBOs needed. ★FBOs most accountable to farmers, but high externalities. ★PFAE outcomes mixed.

Extension commercialization: How much to charge for extension services	Ariel Dinar (1996)	*American Journal of Agricultural Economics*, vol. 78, no. 1	Calculates a supply (marginal cost) function and demand function for price of private extension, while commenting on outcomes of partial privatization.	★Different forms of extension popular in different countries and require different agent/firm structures. ★Commercialization is not equitable to smallholders. ★Commercialization→diversification→price increase.
Contemporary experiences in extension reform: Insights from Pakistan and Mozambique	William M. Rivera (2006)	*Journal of International Agricultural and Extension Education*, vol. 13, no. 1	Is privatization a practical approach to meeting the demand for information by all or some producers? Compares and analyzes results of two case studies (see next cell).	★2002 Workshop advocates privatization, demand-driven programs, decentralization, and revitalization of public sector extension. ★Extension reform should not be advocated for on the basis of ideology (like the Washington Consensus) but based on individual country analysis (aka, "Best Fit" model). ★Thoughtless privatization leaves the poor not only uninformed, but misinformed. ★Impetus for reform should come from farmers for reform to be successful ("demand-driven"). ★Reform must also be environmentally conscious (requiring public sector because of positive externalities).
Global trends in extension privatization	William M. Rivera (1992)	*Journal of Extension*, vol. 30, no. 3	How is extension being reformed worldwide?	★Three options have emerged: public financing by taxpayers only for services of direct concern to general public, the rest from users (France); direct charging of users without privatization (the United Kingdom); 50–50 split of responsibility for providing extension between public and private sector with coordination and research responsibility belonging to the public sector (the Netherlands).

(Continued)

Table 2.3 (Continued)

Article title	Author	Publication information	Issues addressed	Conclusions reached
Linking smallholder farmers to markets and the implications for extension and advisory services	Shaun Ferris et al. (2014)	MEAS Discussion Paper 4	Are certain types of market support more appropriate for specific farmer segments? Which intervention best for smallholders? Improved market access help poor farmers escape poverty?	★Woman and youth key candidates for intervention. ★Farmer and market linkage must be calibrated. ★Farmer agency must be developed. ★Contract farming and certification schemes have been effective. ★Government should foster pluralism. ★Capacity of managers and field agents must be developed. ★ICT, infrastructure, management, and coordination must be developed. ★Collective action is essential.
Impacts of extension privatization	William M. Rivera (1993)	*Journal of Extension*, vol. 31, no. 3	What has been the effect of extension privatization worldwide?	★Privatization has negative effects on small farming. ★Privatization and competition have decreased cooperation between extension agents and clients as well as within the extension industry. ★Privatization ignores environmental problems.
Reconsidering agricultural extension	Robert Picciotto and Jock R. Anderson (1997)	*World Bank Research Observer*, vol. 12, no. 2	Details the causes and symptoms of the shift from top-down, paternalistic extension to the new pluralistic paradigm that developed in the late 1990s.	★The growing influence of New Institutional Economics (especially Umali-Deininger's work), the strained capacities of the public sector, the growing need to provide specialized extension, and the development of new ICTs (as well as the opposite-side pressure of the public sector providing services that result in significant, positive externalities) resulted in the pluralistic paradigm of the late 1990s.

The private sector's role in agricultural extension systems: Potential and limitations	Gershon Feder, Regina Birner, and Jock R. Anderson (2011)	*Journal of Agribusiness in Developing and Emerging Economies*, vol. 1, no. 1	Potential and limitations of private sector in new pluralistic paradigm of extension and in the context of the new, wider agenda of extension; analyzes case studies and suggests some areas for further research.	*New agenda: market linkages, environmental advice, entrepreneurial/business advice, tech transfer, capacity and skill development of farmers and other agents.
				*Most policymakers still view extension as tech transfer.
				*Need a balance between public and private to limit state and market failures inherent in extension.
				*Pure private extension doesn't address poverty-alleviation criteria.
				*Vouchers address state and market failure balance but come with practical problems of fraud.
				*Public sector role in training and organizing trainers.
				*Affirmative action avoids problems of elite capture in FBOs.
				*Well-qualified extension providers scarce in private sector, hard to move them out of public sector.
				*Privatization of extension is less successful if public extension still operate in parallel (Nicaragua and Uganda).
				*Although case studies show that contracting out extension is generally unsuccessful, the success of contracting out other government functions indicates that contracting out extension needs to be properly tailored for success.

(Continued)

Table 2.3 (Continued)

Article title	Author	Publication information	Issues addressed	Conclusions reached
Effectiveness of Tata Kisan Sansar in technology advisory and delivery services in Uttar Pradesh	Anirban Mukherjee et al. (2011)	*Indian Research Journal of Extension Education*, vol. 11, no. 3	Measures effectiveness of private embedded extension provider Tata Kisan Sansar to 50 farmers.	★Effectiveness index composed of following elements: input delivery system, types of service, extent of adoption, increased yield of farmers, increased income of farmers, satisfaction index. ★54% of farmers' input resulted in a medium effectiveness score (41–60) for TKS, whereas the input of the other 54% resulted in a high effectiveness score (61–80).
Final report. Expert consultation on the G8 new alliance for food security and nutrition ICT extension challenge	Julie Howard, Mark Bell, Judith Payne, other ICT Extension Challenge participants (2012)	USAID	Summarizes findings and outcomes of G8 new alliance as well as ICT extension challenge.	★Several specific factors for success of extension are mentioned in paper but the necessity of strong feedback loops, taking advantage of new ICT (although Bell mentioned that it wasn't a silver bullet and human participation was still key), social marketing, prioritizing relevance and applicability, and effective monitoring and measurement were repeated often throughout this summary.
Policy brief: Tapping the energy of farmers' creativity: Supporting farmer-led joint research	Chesha Wettasinha and Ann Waters-Bayer (2010)	Prolinnova International Secretariat	Summarizes Prolinnova's work fostering farmer-led joint research.	★Partners for farmers should be chosen based on type of local innovation being done by farmers. ★FLRJ not only approach, part of pluralistic system. ★Farmers need to be confident and assertive. ★Women should be included. ★Communication key. ★Farmers need access to resources for funding their research.

Title	Author (Year)	Journal	Description	Key points
Extension services and multifunctional agriculture: Lessons learnt from the French and Dutch contexts and approaches	Pierre Labarthe (2009)	*Journal of Environmental Management*, vol. 90, suppl. 2	Analyzes consequences of acknowledgement of the multifunctionality of agriculture and the trend of privatization/ commercialization through and a framework that combines SNA and IEA approaches.	★Room for this new approach needs to be made in the agendas and structure of agricultural research/development systems. ★Privatization of extension will lead to decrease in quality because linkages will weaken and competition inherent in privatization leads to withholding of new information.
Challenges and prospects of privatization of agricultural extension service delivery in Nigeria	P.E. Adejo, O.J. Okwu, and M.K. Ibrahim (2012)	*Journal of Agricultural Extension and Rural Development*, vol. 4, no. 3	Analyzes positives and negatives of whether Nigeria should privatize its extension services (as well as addressing the context of privatization and some issues and methods of implementation).	★Providing credit is crucial to the success of privatization because a large majority of Nigerian farmers are smallholders. ★Institutional structure of Nigerian extension must change to support privatization. ★Farmers and all stakeholders must participate in reform. ★The public sector must withdraw gradually and never absolutely. ★The quality and professionalism of extension services must be ensured. ★Farmers' cooperatives must be strengthened. ★Corrupt must be addressed.

(Continued)

Table 2.3 (Continued)

Article title	Author	Publication information	Issues addressed	Conclusions reached
Privatising agricultural extension: *Caveat emptor*	A.D. Kidd, J.P.A Lamers, P.P. Ficarelli, and V. Hoffman (2000)	*Journal of Rural Studies*, vol. 16, no. 1	Describes forces that create the context of the privatization of extension and examines the roles of the public and private sector in the consequent transition.	*Tone of article is very skeptical about privatization and somewhat pro–public sector (he uses the world "fashionable" to describe the movement toward privatization and market liberalization in general quite frequently). *Farmer association and agricultural advisor development/capacity building should be the responsibility of the public sector because they are public goods. *Reform should be tested through pilot programs whenever possible.
Privatization of agricultural extension services in the EU: Towards a lack of adequate knowledge for small-scale farms?	Pierre Labarthe and Catherine Laurent (2013)	*Food Policy, vol. 38*	Does privatization of agricultural extension in the EU have adverse effects on small farms?	*After WWII small farms slowly vanished from the EU policy agenda until the joining of 12 new member states with significant numbers of smallholders brought the issue back on the agenda in 2011. *Privatization sounded good in theory because outputs analyzed were all from neoclassical economics, which simplified and ignored relevant aspects like scale of farms, transaction costs, and the importance of institutions. *Privatization combined with social exclusion, organizational problems, and misconceptions about extension reform lead to near total exclusion of smallholders.

| Demand driven agricultural advisory services | Sanne Chipeta (2006) | Neuchâtel Group | Why are demand-driven agricultural advisory services beneficial and how should they be fostered? | ★My thought: Neoclassical side is best represented by Umali's good matrix, whereas other approaches (especially, institutional economics) are best represented by Birner's "Best Practice" paper.
★Public sector must enact enabling policies and commit to transition.
★Pluralism of well-qualified providers and farmers being perceived as client not subject are key, but government should not provide free extension that competes with private sector.
★Farmer demand can be influenced by demands on farmers (government regulations, retailer standards, etc.)—this can be used to make agriculture more environmentally friendly.
★Farmers organizations must be strengthened to articulate demand and public funding can be used to help.
★Government should regulate quality standards based on farmers' interests to prevent market failure resulting from asymmetrical information.
★Linkages between research and farmers are vital.
★College agriculture programs should be in line with needs of farmers (especially business skills). |

(Continued)

Table 2.3 (Continued)

Article title	Author	Publication information	Issues addressed	Conclusions reached
New challenges in agricultural advisory services from a research perspective: A literature review, synthesis and research agenda	Guy Faure, Yann Desjeux, and Pierre Gasselin (2012)	*Journal of Agricultural Education and Extension*, vol. 18, no. 5	Literature review of major debates in agricultural advisory services literature from 1998 to 2008 around five themes: (1) Institutional environment of AAS. (2) Structures needed for operation of an advisory system. (3) Actors providing advisory services. (4) Approaches, methods, tools, and content of advisory. (5) Evaluation and impact methodologies of AAS.	★Uses Birner et al. (2009) framework to analyze AAS literature. ★AAS part of broader system (AKIS). ★Significant debate in literature comparing demand-driven to market-driven AAS. ★Theoretically AAS should be diverse and customized. ★Providers need to be mediators or mediators need to become involved (see innovation intermediaries). ★At present AAS is managed by FPOs, supply-chain interprofessional organizations, private providers, or contract farming. ★Most literature points out drawbacks of privatization but does not reject it as an option. ★Public interests must be considered and public regulations must be in place to prevent market failures. ★Advisors should teach management skills as well (soft system). ★Advisory can be divided between soft and hard systems or between technology transfer, advice, and learning facilitation (TT is best for simple problems LF is best for complex). ★FPOs have difficulty exerting influence when they are not involved in the provision of extension.

Title	Author(s)	Journal	Research Question	Key Findings
				★Variety of approaches have pluralistically supplanted TT. ★Birner's "Best Fit" model is best in theory but is difficult to operationalize. ★Co-creation of knowledge is emphasized. ★Ability to acquire new knowledge is a function of age, experience, level of formal education in Nigeria (Okunade, 2007). ★In Benin paying for knowledge is perceived as sinful.
Towards a better conceptual framework for innovation processes in agriculture and rural development: From linear models to systemic approaches	Karlheinz Knickel, Gianluca Brunori, Sigrid Rand, and Jet Proost (2009)	*Journal of Agricultural Education and Extension,* vol. 15, no. 2	How can a better framework be created to address the recognition of the multifunctionality of agriculture by EU CAP and RD documents and the deficiencies of conventional economic approaches.	★CAP and RD policies now look at 3 objectives: rural restructuring (closest link to traditional production-focused policy), environmental concerns, and wider needs of rural areas; with a focus on decentralization and flexibility. ★Extension should focus on turning farmers into rural entrepreneurs while taking a much broader approach and looking territorially instead of sectorally. ★Progressive shift from innovation being perceived as linear, exogenous, and tech transfer oriented to systemic, endogenous, and learning process oriented. ★Social learning is at core of innovation. ★Learning process is composed of socialization, internalization, and recombination. ★Demand-driven model does not converge with public interests→second-order innovation needed.

(Continued)

Table 2.3 (Continued)

Article title	Author	Publication information	Issues addressed	Conclusions reached
Matching demand and supply in the agricultural knowledge infrastructure: Experiences with innovation intermediaries	Laurens Klerkx and Cees Leeuwis (2008)	*Food Policy,* vol. 33, no. 3	How can Innovation Intermediaries solve market and systemic failures on demand and supply sides of agricultural innovation?	★Innovation is not linear so IIs are needed. ★IIs solve problem of articulating demand and response (bridging cognitive distance), bridging managerial gap, and addressing asymmetrical information that causes market failures and misconceptions about agricultural extension. ★Helps farmers obtain credit. ★Helps facilitate the co-creation that is essential to the innovation process. ★See typology of IIs on USB. ★Serves as a cultural and cognitive bridge. ★Develops innovative concepts outside of market and policy pressures. ★Sensitive to regional contexts. ★Functions as a liaison. ★Problem is hard to determine service value. ★Too much focus on entrepreneurial farmers (misses subsistence smallholders). ★Impartiality and neutrality of IIs is critical. ★Public support is need but difficult to make the case for.
Agricultural extension: Good intentions and hard realities	Jock R. Anderson and Gershon Feder (2004)	*World Bank Research Observer,* vol. 19, no. 1	What considerations lead policymakers to see extension as a public responsibility and what factors explain differences	★Difficulties in public extension: managing scale and complexity; policy environment, linkage with knowledge generation, difficulty in attributing impact to extension, weak accountability, distracting public duties, political climate, fiscal sustainability.

Title	Author(s)	Journal	Research questions	Findings
			in performance between different modalities? Develops a conceptual framework based on welfare economics and analyzes differences in efficiency.	★Rejects T&V, considers decentralization with local government delivery or local delivery by farmer associations and fee for service extension or contracting/vouchers and farmer field schools (too expensive).
Private agricultural extension system in Kenya: Practice and policy lessons	Milu Muyanga and T.S. Jayne (2008)	*Journal of Agricultural Education and Extension*, vol. 14, no. 2	Private sector more efficient than public sector in terms of delivery? What circumstances make private extension more efficient than public extension (related to the question of when does extension cease to be a public good)? Are parallel public and private extension systems prudent? What is the new role of the government?	★Private sector cannot replace public sector completely. ★Public sector must either contract private or serve unprofitable areas (complementing not competing with private sector). ★Context: three systems currently in Kenya: (1) public (top-down, uniform, inflexible), (2) commodity-based (only serves profitable commercial farmers), (3) private extension (commercial and noncommercial). ★Most providers are using farmers' groups (cost of seeing one farmer reduced by half when in a group). ★Marketing strategy is an essential factor but is one of the least adopted "good practices." ★Government should provide infrastructure to marginalized areas.

(Continued)

Table 2.3 (Continued)

Article title	Author	Publication information	Issues addressed	Conclusions reached
Agricultural extension policy and practice in Australia: An overview	S.P. Marsh and D.J. Pannel (1999)	*Journal of Agricultural Education and Extension*, vol. 6, no. 2	Analyzes change in extension in Australia after reform based on public/private good distinction as well as other new ideological focus like competitive neutrality.	★Research and development corporations (RDCS)—half public, half private companies—are playing a bigger role as they are becoming more privatized and creating better public/private liaisons. ★To prevent crowding out government should only intervene in public good areas (in terms of provision, not necessarily funding). ★Public funds used to achieve state objectives like environmental conservation, privatization is reducing their ability to do this. ★State agencies have changed to become more accountable and decentralized through several reforms, especially funder–purchaser–provider models (FPP), contracting (all types of goods), and cost recovery (for private goods). ★Extension should be "demand-pull" not "science-push." ★National programs use farmer groups as vehicles to maintain control over extension (reducing duplication, etc.). ★Farmer-driven extension but only within top-down goals, creates conflict of interest. ★Privatization methods: ceasing to provide funds for certain areas, charging for private good info, selling service unit to private sector (de facto in Australia, by policy in New Zealand). ★Growing commitment to participatory processes.

Title	Author	Source	Description	Key Points
Public–private policy change and its influence on the linkage of agricultural research, extension and farmers in Iran	Esmail Karamidehkordi (2013)	Journal of Agricultural Education and Extension, vol. 19, no. 3	Examines research–extension–farmer linkages (REFL) through a comparison of case studies in Iran from the years 1999, 2005, and 2010. Also details a brief ideological history of extension paradigms.	*1960s transfer of technology (TOT) emerges from research, development, and delivery (RD&D) paradigm—in the late 1980s and early 1990s an emphasis on soft systems thinking results in a movement towards privatization. *Attitudinal change from provider–user relationship to stimulator–client relationship. *Privatization changed status of REFL from mediocre to near nonexistent with many extension providers not acknowledging research as a legitimate source of information. *Pluralistic system better than purely private.
Review of agricultural extension in India	Marco Ferroni and Yuan Zhou (2012)	In: Transforming Indian Agriculture – India 2040, SAGE Publications	What is the best way to provide meaningful advice and create learning environments that can help achieve desired outcomes? Reviews issues and performance of extension from primary production and linkage/value chain points of view.	*"Extension is a conveyor belt" (shows that author views extension as a linear model). *Details history emphasizing T&V of 1970s and ATMA pre and post reform. *Farmer participation, innovations systems, market-based, and demand-driven are emphasized trends. *T&V and ATMA both intended as coordinating frameworks. *ATMA not silver bullet because of implementation problems. Access is most pervasive issue with public extension. *Public extension more dynamic in commercially vibrant areas. *Research–extension link is weak. *MANAGE offer DAESI (diploma for input dealers to provide extension). *MKV learn by doing approach successful (may be because they also provide quality inputs). *Contract farming unites front end and back end of extension. *Public–private partnerships are tool of choice.

(Continued)

Table 2.3 (Continued)

Article title	Author	Publication information	Issues addressed	Conclusions reached
Agricultural extension, collective action and innovation systems: Lessons on network brokering from Peru and Mexico	Jon Hellin (2012)	*Journal of Agricultural Education and Extension*, vol. 18, no. 2	Examines research in Peru and Mexico that explores new approaches to extension delivery.	*Farmers' associations are essential to overcoming the many barriers associated with being a marginalized farmer but are difficult to form because of cultural reasons among other reasons. *Extension agents (especially NGOs) make good network brokers. *Kamayoq model demonstrates that fee-for-service extension is viable for marginalized areas/groups. *Concurs that there is no ideal "one-size-fits-all" approach. *Concurs that public–private partnerships are better than pure versions of one or the other. *Whether extension contributes to fostering collective action and AISs is more important than who funds/provides the extension. *Also, equally important are strengthening the capacities of actors, and creating networks between heterogeneous actors.
Farmer experience of pluralistic agricultural extension, Malawi	Clodina Chowa, Chris Garfoth, and Sarah Cardey (2013)	*Journal of Agricultural Education and Extension*, vol. 19, no. 2	Has extension reform to support pluralism in Malawi led to more demand-driven extension.	*Farmers like having access to a variety of information and technology, but providers are still acting paternalistically toward farmers and not incorporating their needs. *Farmer groups exist but must be strengthened through extension. *Lack of linkages between input, markets, and farmers.

| | | | | *Local institutions (especially local governments) can broker interactions between providers and farmers to promote coverage and responsiveness.
*See paper for brief history of extension in Malawi.
*Innovation system constructs must be farmer-centric to help ensure demand-driven extension.
*Microfinance and social service providers can indirectly provide agricultural help.
*Radio listening groups lead to farmers seeking out fee-for-service extension.
*Providers need to coordinate so as not to promote redundancy and not to confuse farmers and deter them from acting (need institutional regulation, possibly through government).
*Systemic failure in reform but it is improvable through policy intervention. |
| Contract farming in developing countries: Patterns, impact, and policy implications | Nicholas Minot (2007) | Case Study 6-3, Food Policy for Developing Countries: The Role of Government in the Global Food System | What is the role and impact of contract farming, especially in helping small farmers, and how can its growth be facilitated by policy. | *Empirical studies agree that it improves the income of farmers.
*Extension (especially private) and contract farming are mutually reinforcing and this reinforcement should be supported by the policy environment.
*Intermediaries (especially extension agents) are important for facilitating contracts and preventing contract violation.
*Policy environment should promote public–private partnerships in extension.
*Extension must provide marketing assistance in addition to the usual production-focused advice. |

(Continued)

Table 2.3 (Continued)

Article title	Author	Publication information	Issues addressed	Conclusions reached
The impact of the national agricultural advisory services program on household production and welfare in Uganda	Geofrey Okoboi, Annette Kuteesa, and Mildred Barungi (2013)	Brookings' Africa Growth Initiative, Working Paper 7	Examines level of participation of vulnerable households (headed by women, youths, or people living with disabilities) and impact on accessibility and several impact of NAADS.	★Vulnerable households participated less. ★NAADS increased accessibility to extension services and credit regardless of vulnerability, but quality of services was suspect. ★Vulnerable households should be targeted through affirmative action. ★Farmers' capacity must be built.
Capitalisation of Samriddhi's Experiences on Private Rural Service Provider System	H. Martin Dietz, Noor Akter Naher, and Zenebe Bashaw Uraguchi (2013)	SDC – HELVETAS Swiss Inter-cooperation, Bangladesh	Examines case study in Bangladesh for potential model of private extension service.	★LSPs (both individually and when organized as SPAs) act as extension workers (independently or for an output actor), and/or help organize farmers into producer groups. ★LSPs work part time and full time. ★Providers receive training from government line agencies, input actors, and output actors. ★Good linkages with other entities essential.
Systemic evaluation of advisory services to family farms in West Africa	Guy Faure, Pierre Rebuffel and Dominique Violas (2011)	*Journal of Agricultural Education and Extension*, vol. 17, no. 4	Develops a conceptual framework based on systems theory to analyze how advisory services can better address the needs of family farm providers.	★Budget and human resource capabilities are important to keep in mind. ★Governance mechanisms (especially funding mechanism if public–private partnership) have a strong influence on quality of linkages. ★Extension cannot be reduced to demand or supply driven but must be based on interactions between farmers and advisors.

Title	Author (Year)	Journal	Aim	Key points
Extension in tough times: Addressing failures in public and private extension, lessons from the Tasmanian wool industry, Australia	Warren Hunt and Jeff Coutts (2009)	*Journal of Agricultural Education and Extension*, vol. 15, no. 1	Reports research on impact of a range of extension approaches on the Tasmanian wool industry in Australia.	★Traditional public and private extension approaches (modeled after Umali's good matrix) failed, leaving a void. ★A wool growers' group funded the Tasmanian Institute of Agricultural Research (TIAR) (an arm of the University of Tasmania) to provide extension. ★Impact evaluated using Bennett's hierarchy. ★Majority improved technical, management, and information seeking skills. ★Audience segmentation approach worked (echoing "one size does not fit all").
The state of agricultural extension: An overview and new caveats for the future	Amanda Benson and Tahseen Jafry (2013)	*Journal of Agricultural Education and Extension*, vol. 19, no. 4	Aims to provide a global overview of agricultural extension changes.	★Macroeconomic circumstances of late 1970s/ early 1980s (interest rate rise followed by recession) led to price of agricultural commodities falling leading to a surplus of agricultural goods (especially in northern countries), leading to a decrease in public spending and to market deregulation and privatization. ★Extension moved from production focused to market oriented. ★Improved monitoring, evaluation, and knowledge sharing/linkages emphasized in the literature. ★Privatization decisions should not be based on narrow experiences in northern countries (especially when applied to the South). ★Private sector not just in delivery, private sector funds equal amount of research as public sector (50–50). ★Collective action by small farmers takes advantage of economies of scale.

(Continued)

Table 2.3 (Continued)

Article title	Author	Publication information	Issues addressed	Conclusions reached
				★Collective action is difficult for the poorest. ★Literature agrees that when extension deals with societal good or farmers' lack information to decide whether or not to pay for extension, government should step in. ★Decentralization eliminates economies of scale. ★Pluralism is good but difficulties in maintaining it arise when the public sector views the private sector as a rival and when there is confusion over division of roles.
Operationalizing demand-driven agricultural research: Institutional influences in a public and private system of research planning in the Netherlands	Laurens Klerkx and Cees Leeuwis (2009)	*Journal of Agricultural Education and Extension,* vol. 15, no. 2	Examines the practical considerations of carrying out demand-driven research and extension.	★Farmers should have a role in planning, but it is difficult to ensure that they do. ★Brokers and facilitators needed to match demand to supply.
Improving extension programs: Putting public value stories and statements to work	Nancy K. Franz (2013)	*Journal of Extension,* vol. 51, no. 3, Article 3TOT1	Aims to apply public good movement to extension work.	★Public value movement from public administration has spread to extension because of external accountability requirements. ★Works should create public value stories and statements to enhance effectiveness of extension and increase the transformative impact of their programs. ★"Transformative education" should be emphasized.

Title	Author	Source	Purpose	Notes
Privatization lessons for U.S. extension from New Zealand and Tasmania	Peter Bloome (1993)	*Journal of Extension*, vol. 31, no. 1, Article 1INTL1	Seeks to find how case studies in Tasmania, Australia, and New Zealand are applicable to U.S. extension reform.	★Tasmania, Australia: government agency became a state owned enterprise. ★Australia and New Zealand governmental extension services emphasized private good components of extension, therefore privatization was easier. ★Cost recovery failed. ★Poor implementation because government agency wasn't able to make the organizational change from the public to private sector. ★Public benefits of extension couldn't be commercialized. ★Bloome concludes public extension can't be replaced through privatization.
What does it mean to be socially responsible? Case study on the impact of the producer–plus program on communities, women, and the environment in Ecuador	Trent Blare and Pilar Useche (2014)	MEAS Case Study 11	Analyzes Transmar Ecuador's effort to provide extension to cacao farmers.	★Difficult to include women and marginalized groups/areas. ★Farmers' cooperatives are important to solve problem of reaching marginalized groups. ★Jointly markets and delivers extension to farmers. ★A variation of embedded extension (similar to contract farming). ★Transmar gets rid of middleman to stop exploitation of smallholders and ensure that they get a larger piece of the value chain (positive incentives). ★Vertical integration—leads to quality control to meet demands of consumers. ★Ecuadorian debt crisis and consumer demands created opportunity for Transmar. ★E-payment system makes it hard to reach the poorest farmers.

(Continued)

Table 2.3 (Continued)

Article title	Author	Publication information	Issues addressed	Conclusions reached
Research findings: The role and function of agricultural extension	A. Blum, A. Lowengart-Aycicegi, and H. Magen (2010)	International Potash Institute, Research Findings: e-ifc No. 25	Presents evidence from government efforts to decentralize and privatize agricultural extension services.	★Four types of private advisors: input, output, marketing chains, extension business actors. ★Many projects collapse when foreign donors leave—to solve budget should be affordable, infrastructure should be developed, all stakeholders should be involved throughout the process. ★Meta-studies suggest that rate of return on agricultural extension investment is high. ★Sociological impact indicators necessary to complement economic indicators.
Achievements and challenges in agricultural extension in India	Marco Ferroni and Yuan Zhou (2012)	*Global Journal of Emerging Market Economies*, vol. 4, no. 3	Builds on discussion of new status of extension, focusing specifically on India.	★Mobile applications have significant potential to widely disseminate agricultural advice to fill knowledge gaps. ★NGOs and commercial organizations are growing and thriving and should keep doing what they are doing. ★Input dealers and progressive farmers still first destination for agricultural advice. ★Small farmer coverage and clarity of role of public sector are two main unaddressed issues. ★Public private partnerships are key.
ICTs: Transforming agricultural extension? Report of the 6th Consultative Expert Meeting of CTA's Observatory on ICTs	Don Richardson (2006)	CTA Working Document Number 8034	Analyzes role of ICT in extension, especially in relation to what CTA (ACP-EU Technical Centre for Agricultural and Rural Cooperation) can do.	★Extension changing from transfer of technology paradigm to facilitation, brokerage, and communication paradigm. ★Extension must focus and be analyzed on all aspects of rural livelihoods (*note*: similar to the MFA—multifunctionality of agriculture concept). ★Connectivity constraints still an issue for using ICT in extension.

Title	Author	Source	Description	Notes
Evaluating the impact of agricultural extension programmes in sub-Saharan Africa: Challenges and prospects	Hailemichael Taye (2013)	*African Evaluation Journal*, vol. 1, no. 1	*Analyzes impact evaluation studies in sub-Saharan Africa (SSA) to ascertain problems with reports and look at future prospects.	*CTA's role is to clarify, facilitate development of, and spread ICT knowledge in relation to extension, incorporate multiple–stakeholder planning approaches, and develop steps for overcoming bottlenecks. *Most impact evaluations overestimate because of poor evaluation methodologies, lack of reliable data, and/or insufficient capacity to conduct rigorous evaluations. *Focus should be on improving impact not proving impact. *New focus should also be put on innovating monitoring, evaluation, and learning tools that acknowledge the dynamic nature of agricultural development.
How to promote institutional reforms in the agricultural sector? A case study of Uganda's national agricultural advisory services (NAADS)	Patience, B. Rwamigisa, Regina, Birner, Margaret, N. Mangheni and Arseni Semana (2013)	Conference paper presented at the International Conference on the "Political Economy of Agricultural Policy in Africa," 18–20 March 2013, Pretoria, South Africa	Seeks to explain limited success of NAADS program in Uganda.	*Reform process was shaped by interaction between donor dominated coalition and technically oriented coalition led by government (representing radical vs. gradual change). *Exclusion of gradual coalition led to early political capture and consequent implementation/governance problems. *Solution is consensus building and gradual approach to new reforms.
Viewing Bennett's hierarchy from a different lens: Implications for extension program evaluation	Rama Radhakrishna and Cathy F. Bowen (2010)	*Journal of Extension*, vol. 48, no. 6, Article 6TOT1	Describes use of Bennett's Hierarchy of Change in extension program evaluation in the last 35 years.	*Seven steps: 1–4 focus on process evaluation, 5–7 focus on outcome/impact evaluation. *Evaluation should be considered during planning phase, not just post-program. *Useful to show accountability to stakeholders. *Some steps require complex indicators but most only require simple and cheap indicators.

(*Continued*)

Table 2.3 (Continued)

Article title	Author	Publication information	Issues addressed	Conclusions reached
Privatization of agricultural extension	R. Saravanan (2001)	In: Private Extension in India: Myths, Realities, Apprehensions and Approaches. National Institute of Agricultural Extension Management (MANAGE)	Describes reasons behind privatization of agricultural extension and makes suggestions for the future.	*Reasons behind privatization: financial burden on government, poor public extension performance, commercialization, and need for specialization in agriculture. *Private extension mostly focuses on big farmers (Saravanan goes on to suggest a similar division of labor for public–private partnerships as Umali, with the public sector serving marginalized, unprofitable farmers and delivering public goods, and the private sector serves the commercial farmers and maximizes farmers). *Complex, demand–driven technology transfer should be privatized first as it is the easiest to commercialize. *Contract and sharecropping systems should be implemented. *Farmer cooperatives should be strengthened. *Public sector should withdraw gradually.
Agricultural extension in India: Current status and ways forward	Sulaiman V. Rasheed (2012)	Background paper prepared for Roundtable Consultation on Agricultural Extension, Beijing, China, March 15–17, 2012	Seeks to review the evolving landscape of Indian agricultural extension and looks to the future.	*Innovation systems approach taking hold—emphasizing communication in a nonlinear context (especially in relation to network building, supporting social learning, and dealing with dynamics of power and conflict). *Input: extension can be viewed as a facet of marketing—Nuziveedu sees has invested heavily in extension for marketing their seeds.

* Agribusiness firms (output): Indian Tobacco Company's (ITC) e-choupal kiosks run by a local farmer (also an example of ICT extension) as well as choupal saagar hypermarkets for farmers, PepsiCo implementing high quality seeds programs.
* Farmer Cooperatives: Grape Growers Association of Maharashtra conducts group discussion as well as R&D.
* NGO: Syngenta Foundation promoting sustainable agricultural practices.
* ICT: e-choupals and Ms Swaminathan Research Foundation, Digital Green (which fields a website and human mediation).
* Financial institutions: National Bank for Agriculture and Rural Development (NABARD).
* Consultancy: variety of fee-for-service providers.
* India should support: pluralism and partnerships, coordination and more funding, research aimed at smallholders, and change management.

REFERENCES

Adejo, P.E., Okwu, O.J., Ibrahim, M.K., 2012. Challenges and prospects of privatization of agricultural extension service delivery in Nigeria. J. Agric. Ext. Rural Dev. 4 (3), 63–68.

Anderson, J.R., Feder, G., 2004. Agricultural extension: good intentions and hard realities. World Bank Res. Obs. 19 (1), 41–60.

Babu, S.C., Joshi, P.K., Glendenning, C., Asenso-Okyere, K., Rasheed Sulaiman, V., 2013. The state of agricultural extension reforms in India: strategic priorities and policy options. Agric. Econ. Res. Rev. 26 (2).

Benson, A., Jafry, T., 2013. The state of agricultural extension: an overview and new caveats for the future. J. Agric. Educ. Ext. 19 (4), 381–393.

Birner, R., Anderson, J.R., 2007. How to Make Agricultural Extension Demand-Driven? The Case of India's Agricultural Policy. International Food Policy Research Institute (IFPRI), Washington, DC. Discussion Paper 00729.

Birner, R., Davis, K., Pender, J., Nkonya, E., Anandajayasekeram, P., Ekboir, J., et al., 2009. From "best practice" to "best fit": A framework for designing and analyzing pluralistic agricultural advisory services worldwide. J. Agric. Educ. Ext. 15 (4), 341–355.

Blare, T., Useche, P., 2014. What Does it Mean To Be Socially Responsible? Case Study on the Impact of the Producer-Plus-Program on Communities, Women, and Environment in Ecuador. MEAS Case Study 11. Modernizing Extension and Advisory Services, Urbana, IL.

Bloome, P.D., 1993. Privitization lessons for U.S. extension from New Zealand and Tasmania. J. Ext. 31 (1), Article 1INTL1. Available at www.joe.org/joe/1993spring/intl1.php.

Blum, A., Lowengart-Aycicegi, A., Magen, H., 2010. The role and function of agricultural extension. International Potash Institute, Research Findings: e-ifc No. 25. Available at www.ipipotash.org/en/eifc/2010/25/2.

Chipeta, S., 2006. Demand Driven Agricultural Advisory Services. Neuchâtel Group.

Chowa, C., Garforth, C., Cardey, S., 2013. Farmer experience of pluralistic agricultural extension. Malawi. J. Agric. Educ. Ext. 19 (2), 147–166.

Dinar, A., 1996. Extension commercialization: How much to charge for extension services. Am. J. Agric. Econ. 78 (1), 1–12.

Dinar, A., Karagiannis, G., Tzouvelekas, V., 2007. Evaluating the impact of agricultural extension on farms' performance in Crete: a nonneutral stochastic frontier approach. Agric. Econ. 36 (2), 135–146.

Dietz, H.M., Naher, N.A., Uraguchi, Z.B., 2013. Capitalisation of Samriddhi's experiences on private rural service provider system. SDC – HELVETAS Swiss Intercooperation, Bangladesh.

Faure, G., Rebuffel, P., Violas, D., 2011. Systemic evaluation of advisory services to family farms in West Africa. J. Agric. Ext. Educ. 17 (4), 325–359.

Faure, G., Desjeux, Y., Gasselin, P., 2012. New challenges in agricultural advisory services from a research perspective: a literature review, synthesis and research agenda. J. Agric. Educ. Ext. 18 (5), 461–492.

Feder, G., Birner, R., Anderson, J.R., 2011. The private sector's role in agricultural extension systems: potential and limitations. J. Agribusiness Dev. Emerg. Econ. 1 (1), 31–54.

Ferroni, M., Zhou, Y., 2012. Review of agricultural extension in India. In: Transforming Indian Agriculture – India 2040: Productivity, Markets, and Institutions. SAGE Publications, Washington DC, pp. 187–242.

Ferris, S., Robbins, P., Best, R., Seville, D., Buxton, A., Shriver, J., et al., 2014. Linking Smallholder Farmers to Markets and the Implications for Extension and Advisory Services. MEAS Discussion Paper 4. Modernizing Extension and Advisory Services, Urbana, IL.

Ferroni, M., Zhou, Y., 2012. Achievements and challenges in agricultural extension in India. Global J. Emerg. Mark. Econ. 4 (3), 319–346.

Franz, N.K., 2013. Improving extension programs: Putting public value stories and statements to work. J. Ext. 51 (3) Article 3TOT1. Available at www.joe.org/joe/2013june/tt1.php.

Heemskerk, W., Davis, K., 2012. Agricultural Innovation Systems: An Investment Source Book. World Bank Publications, 194–203.

Hellin, J., 2012. Agricultural extension, collective action and innovation systems: lessons on network brokering from Peru and Mexico. J. Agric. Educ. Ext. 18 (2), 141–159.

Howard, P.N., Duffy, A., Freelon, D., Hussain, M., Mari, W., Mazaid, M., 2011. Opening Closed Regimes: What was the Role of Social Media During the Arab Spring? Project on Information Technology and Political Islam. Working Paper 2011.1. University of Washington.

Howard, J., Bell, M., Payne, J., et al., 2012. Final report. Expert consultation on the G8 new alliance for food security and nutrition ICT extension challenge. USAID, Washington DC.

Hunt, W., Courts, J., 2009. Extension in tough times: Addressing failures in public and private extension, lessons from the Tasmanian wool industry. Aust. J. Agric. Educ. Ext. 15 (1), 39–55.

Karamidehkordi, E., 2013. Public–private policy change and its influence on the linkage of agricultural research, extension and farmers in Iran. J. Agric. Educ. Ext. 19 (3), 237–255.

Kidd, A.D., Lamers, J.P.A., Ficarelli, P.P., Hoffmann, V., 2000. Privatising agricultural extension: caveat emptor. J. Rural Stud. 16 (1), 95–102.

Klerkx, L., Leeuwis, C., 2008. Matching demand and supply in the agricultural knowledge infrastructure: experiences with innovation intermediaries. Food Policy 33 (3), 260–276.

Klerkx, L., Leeuwis, C., 2009. Operationalizing demand-driven agricultural research: Institutional influences in a public and private system of research planning in The Netherlands. J. Agric. Educ. Ext. 15 (2), 161–175.

Knickel, K., Brunori, G., Rand, S., Proost, J., 2009. Towards a better conceptual framework for innovation processes in agriculture and rural development: from linear models to systemic approaches. J. Agric. Educ. Ext. 15 (2), 131–146.

Labarthe, P., 2009. Extension services and multifunctional agriculture. Lessons learnt from the French and Dutch contexts and approaches. J. Environ. Manage., 90 (2), S193–S202.

Labarthe, P., Laurent, C., 2013. Privatization of agricultural extension services in the EU: towards a lack of adequate knowledge for small-scale farms? Food Policy 38, 240–252.

Marsh, S.P., Pannel, D.J., 1999. Agricultural extension policy and practice in Australia: an overview. J. Agric. Educ. Ext. 6 (2), 83–91.

Minot, N., 2007. Case Study #6-3, Contract Farming in Developing Countries: Patterns, Impact, and Policy Implications. In: Pinstrup-Andersen, P., Cheng, F. (Eds.), Food Policy for Developing Countries: Case Studies. 13 pp.

Mukherjee, A., Bahal, R., Burman, R.R., Dubey, S.K., Jha, G.K., 2011. Effectiveness of Tata Kisan Sansar in technology advisory and delivery services in Uttar Pradesh. Indian Res. J. Ext. Educ. 11, 8–13.

Muyanga, M., Jayne, T.S., 2008. Private agricultural extension systems in Kenya: practice and policy lessons. J. Agric. Educ. Ext. 14 (2), 111–124.

OECD, 1991. Development Assistance Committee: Principles for the Evaluation of Development Assistance. Organisation for Economic Co-operation and Development, Paris.

OECD, 2011. Monitoring and Evaluation for Adaptation: Lessons from Development Cooperation Agencies, OECD Environment Working Papers No. 38. Organisation for Economic Co-operation and Development, Paris.

Okoboi, G., Kuteesa, A., Barungi, M., 2013. The impact of the national agricultural advisory services program on household production and welfare in Uganda. Brookings' Africa Growth Initiative, Working Paper 7.

Picciotto, R., Anderson, J.R., 1997. Reconsidering agricultural extension. World Bank Res. Obs. 12 (2), 249–259.

Radhakrishna, R., Bowen, C.F., 2010. Viewing Bennett's hierarchy from a different lens: Implications for extension program evaluation. J. Ext. 48 (6) Article 6TOT1. Available at www.joe.org/joe/2010december/tt1.php.

Richardson, D., 2006. ICTs—Transforming Agricultural Extension? Report of the 6th Consultative Expert Meeting of CTA's Observatory on ICTs. CTA Working Document no. 8034. Technical Centre for Agricultural and Rural Cooperation (CTA), Wageningen.

Rivera, W.M., 1992. Global trends in extension privatization. J. Ext. 30 (3), Article 3INTL1. <http://www.joe.org/joe/1992fall/intl1.php>.

Rivera, W.M., 1993. Impacts of extension privatization. J. Ext. 31 (3), Article 3INTL1. <http://www.joe.org/joe/1993fall/intl1.php>.

Rivera, W.M., 2006. Contemporary experiences in extension reform: insights from Pakistan and Mozambique. J. Int. Agric. Ext. Educ. 13 (1), 83–89.

Rivera, W.M., Alex, G., 2004. Privatization of Extension Systems: Case Studies of International Initiatives. Discussion Paper 9. World Bank Publications, Washington, DC.

Rivera, W.M., Rasheed Sulaiman, V., 2009. Extension: object of reform, engine for innovation. Outlook Agric. 38 (3), 267–273.

Rivera, W.M., Qamar, M.K., Van Crowder, L., 2002. Agriculture and Rural Extension Worldwide: Options for Institutional Reform in Developing Countries. FAO (Food and Agriculture Organization of the United Nations), Rome, Italy.

Rwamigisa, B., Birner, R., Mangheni, M., Arseni Semana, A., 2013. How to Promote Institutional Reforms in the Agricultural Sector? A Case Study of Uganda's National Agricultural Advisory Services (NAADS). Paper presented at the International Conference on the Political Economy of Agricultural Policy in Africa. Pretoria, 20–18 March 2013, organized by the Futures Agriculture Consortium and the Institute for Poverty, Land and Agrarian Studies (PLAAS).

Salifu, A., Francesconi, G.N., Kolavalli, S., 2010. "A Review of Collective Action in Rural Ghana", IFPRI Discussion Paper 00998. International Food Policy Research Institute, Washington, DC.

Saravanan, R., 2001. Privatization of agricultural extension. In: Chandra Shekara, P. (Ed.), Private Extension in India: Myths, Realities, Apprehensions and Approaches. Published by National Institute of Agricultural Extension Management (MANAGE), Hyderabad, India, pp. 60–71.

Sulaiman, V.R., 2012. Agricultural Extension in India: Current Status and Ways Forward. Background Paper prepared for the Roundtable Consultation on Agricultural Extension, Beijing, March 15–17, 2012. Centre for Research on Innovation and Science Policy (CRISP), Hyderbad, India.

Taye, H., 2013. Evaluating the impact of agricultural extension programmes in sub-Saharan Africa: Challenges and prospects. Afr. Eval. J. 1 (1).

Umali, D.L., Schwartz, L., 1994. Public and Private Agricultural Extension: Beyond Traditional Frontiers. World Bank Publications No. WDP236, Washington, DC.

Umali-Deininger, D., 1997. Public and private agricultural extension: Partners or rivals? World Bank Res. Obs. 12 (2), 203–224.

Wettasinha, C., Waters-Bayer, A., 2010. Policy brief: Tapping the energy of farmers' creativity: Supporting farmer-led joint research. Prolinnova International Secretariat, The Netherlands.

William Rivera, W., Alex, G., 2004. Extension reform for rural development, Volume 2. Privatization of Extension Systems. World Bank Publications. Agriculture and Rural Development Discussion Paper 9.

CHAPTER 3

Private Approaches to Extension and Advisory Services: A Historical Analysis

Raj Saravanan[1] and Suresh Chandra Babu[2]

[1]Agricultural Extension and Rural Sociology, Central Agricultural University, Pasighat, Arunachal Pradesh, India
[2]International Food Policy Research Institute (IFPRI), Washington, DC, USA

Contents

INTRODUCTION

Extension services provided by private entities working directly with farmers and farming communities have become a key and integral part of emerging pluralistic extension systems in developing countries. However, private approaches in agricultural extension and advisory services have existed for a long time. Globally, voluntary action in agricultural development have been part of the historical legacy. Numerous voluntary and nongovernmental efforts have evolved in the field of agriculture over the centuries. Similarly, Christian and other religious and social service organizations, enlightened princes, individual philanthropists, and social reformers have initiated projects in agricultural development by initiating extension and advisory services. Private extension attempts have been sporadic, operating in smaller areas and with limited impact in the agricultural sector; hence, the history of those attempts has not been

Knowledge Driven Development.
DOI: http://dx.doi.org/10.1016/B978-0-12-802231-3.00003-6

recorded in a systematic way. However, over the last five decades, the emergence of agricultural input companies and nongovernmental organizations in the area of agricultural extension and advisory services has resulted in high visibility for the role of the private sector. The recent emergence of new information and communication technologies (ICTs) has added value to private extension approaches and enabled the entry of several private extension operators.

In this chapter we describe the historical background of the concept of private extension in order to help the reader to understand the context of the case studies analyzed in the rest of this book. In the next section, we provide a clear definition of private extension. This is followed by a section that describes the set of factors contributing to trends in the development of the private extension services.

WHAT IS PRIVATE EXTENSION?

The definition of private extension, seen as a set of services provided by private operators who deal directly with the farmers to address their needs, has evolved over the years. According to Bloome (1993), private extension involves personnel in the private sector who deliver advisory services in the area of agriculture as an alternative to public extension, whereas Van den Ban and Hawkins (1996) argue that in a private setting, farmers are expected to share responsibility for extension services and pay all or part of the cost.

Saravanan and Gowda (1999) operationalized private extension in the following manner in the developing country context: They see private extension as a set of services rendered in the area of agriculture and allied aspects by extension personnel working for private agencies or organizations and for which farmers may or may not be expected to pay a fee; it can be viewed as supplementary or as an alternative to public extension services. Simply put, the process of funding and delivering extension services by private individuals or organizations is called private extension (Chandrashekara, 2001).

In general, however, the concept of private extension emphasizes three aspects: (1) the involvement of extension personnel from private agencies or organizations; (2) payment for services by the clientele through a service fee, directly or indirectly; and (3) playing a supplementary or alternative role to public extension service.

TRENDS IN THE DEVELOPMENT OF PRIVATE EXTENSION AND ADVISORY SERVICES

Traditionally, extension services have been seen as a public good provided by the public sector or the state, as farmers have no resources to pay for information. Further, the sources of information and quality control to ensure that correct information is passed on to the farmers lead the public sector to invest in extension along with research and development. However, over the years several factors have contributed to the decline of public extension, and a pluralistic form of extension has been on the rise. We describe some of these trends below.

Declining budget allocation for public extension by governments and donors

In the last three decades, structural adjustment and stabilization policies promoted by donor communities have resulted in a serious decline in funding for extension from public sources. As a result of financial concerns, many countries have examined alternative structural arrangements, including the feasibility of reducing public sector extension expenditures, enabling changes in agricultural tax rates, charging fees for government extension services, and commercialization and privatization of extension services (Howell, 1985). A steady decrease in the farm population (which is associated with its declining political leverage) is one of the explanations for reductions in public financing for extension and a shift to self-supported or private extension services (LeGouis, 1991). Dinar (1996) documented declining trends since mid-1980s in expenditure for extension in China, Japan, Israel, India, and the United States. The decline in extension expenditure in these and many other countries has been accompanied by structural and institutional changes such as decentralization, commercialization, and privatization. The budget deficits of many national governments make it difficult to pay for extension service (Van den Ban and Hawkins, 1996). Given the considerable decline in budgetary support to agriculture in general and public-sector extension in particular, several countries around the world have experimented with alternative approaches. The partial recovery of costs by the public extension system and privatization of extension services are commonly recommended, as along with experimental alternatives.

Transformation of agriculture from subsistence level to commercialized agribusiness

Another trend that has contributed to the development of private extension as an alternative to public extension is the transformational approach to agriculture. Transforming traditional agriculture from subsistence orientation to the small yet viable agribusinesses has necessitated a role for private extension. This is mainly due to increased knowledge intensity in the agribusiness approach to farming. In the process of agricultural transformation in both developed and developing countries from subsistence level to commercialized agribusiness, the increased investments being made by the private sector in agricultural research, as well as the exclusive rights legally granted to producers of specific technologies as part of the new patent regime, have drawn the attention of the private sector to the area of extension services (Sulaiman and Gadewar, 1994). This trend is also increasing in developing countries, where high-value agriculture has been growing and expanding.

Concerns about efficiency and effectiveness of public extension

The public extension system has been seen as less than efficient, often even ineffective, in connecting to farmers' real need for information. This is due partly to poor accountability of extension workers to the farmers and partly to poor governance and management of the extension services by the public sector. Berdegué (1997) noted that there is no doubt that privatization is significantly more cost-effective than a public alternative would be, that a larger fraction of the total cost can be used to pay for field-level activities rather than for fixed office expenses, and that through a number of progressive improvements the system has become much more cost-effective. Rivera and Cary (1997) noted that the argument for privatization is based upon more efficient delivery of services, lowered government expenditures, and higher quality of services.

Further, the public extension system is unable to cover the vast number of farmers who need extension services in a country. Gustafson (1991) argued that public-sector extension alone will never meet the entire demand for extension services by the world's farmers. The increasing cost of public extension as well as high inefficiency attributed to the government-controlled bureaucratic extension system are the most important stimuli that have speeded up the search for alternative approaches to extension (Sulaiman and Gadewar, 1994). In developing countries in particular, farmers have serious doubts about the role of the public provision of extension. The closeness of politicians to extension workers and the capture of extension benefits by rich and large farmers are often seen as the major sources of distrust between the farmers and the extension systems. It remains a challenge for extension agents employed by the public sector to gain farmers' confidence, without which they have little impact among farmers (Van den Ban and Hawkins, 1996).

Beneficiary contribution for ensuring demand-driven and accountable extension

The increasing cost of public extension and the related inefficiency that comes with provision of "free services" have prompted some authors to call for commitment from the beneficiaries. Antholt (1994) argued that beneficiaries should bear at least some of the cost of services for the following three reasons: First, cost sharing gives farmers ownership and drawing rights. Second, it takes some financial pressure off the government, thereby increasing the chances of sustainability. Third, it provides the basis for a more demand-driven, responsive, and accountable service. Umali and Schwartz (1994) suggested that if customers pay for these services, there is a high likelihood that the provided information will conform to the needs of the farmers, which may make the extension services more cost-effective. Dinar (1996) noted that a total shift from public funding to client funding may not be in the public interest, given the external benefits of technology diffusion and legitimate equity concerns. But there are obvious benefits to "value for money" that are often associated with a demand-driven approach to extension.

In the process of privatization of governmental extension services, it is hoped that by making extension agents accountable to farmers who are competent to judge the quality of their work, this in turn will make the extension service more demand driven, particularly when the farmers can pay for the services they receive. Further, farmers as economic entities, being the main beneficiaries of extension service activities, should be asked to pay for the services they receive as a matter of fairness to the rest of the society (Van den Ban and Hawkins, 1996).

Private extension to supplement and complement public extension

Private extension has always coexisted with public extension, particularly when an input is supplied through a private agent. As a country moves toward easing the public sector away from seed and input provision, there is already space created for the private sector to operate. Then it becomes a question of how the private sector can operate along with public extension and play a meaningful role in increasing farmers' productivity. Wilson (1991) argued that in several developing countries, public–private extension coordination is already established. Alternative patterns include fostering of private corporate initiative, encouraging cooperative ventures by farmers, and privatization of the public extension system. Centralized main line extension services must continue to evolve into a variety of hybrid solutions, combining public support with private delivery methods (Antholt, 1994). For example, cost sharing and a voucher system can increase the voice of farmers in the management of extension services even within the context of public extension.

In the context of the coexistence of public and private extension, Schwartz (1994) argued that the existence of multiple information sources is an advantage for farmers in that they can select the information mix most suited to their goals as producers, and seek the most reliable information sources. Matanmi and Ladele (1996) concluded that if it is well coordinated and given some encouragement, the private sector may significantly complement the efforts of public extension services, thereby moving developing countries' agricultural extension toward sustainability. Some authors have argued for public extension organizations to work toward a certain degree of privatization. For example, Picciotto and Anderson (1997) stated that privatization would allow the public sector to concentrate its limited resources on providing services to neglected areas and high-leverage actions directed at education and training, information technology, and the creation of an enabling framework for equitable and environmentally sustainable rural development. Rivera (2001) noted that in developing countries, pluralism becomes the government's goal, which involves inclusion of other, usually private organizations in both the funding and delivery of extension services.

Decentralization of extension and private extension

A trend that has helped in the development of private extension is the decentralization of service delivery in the public sector. This has implications for how the private sector operates in a region of the country. Rivera (2001) indicated that over the last two decades of the twentieth century, with accelerated agricultural modernization and a macroeconomic reduction of public services, agricultural extension underwent systemic reform and decentralization. This mainly happened through three ways: (1) shifting partial or full authority for extension to lower levels of government or to private entities (structural); (2) through cost sharing and cost recovery schemes (financial); and (3) by democratization of the decision-making process. How such decentralization benefits or hampers the development of the private extension has not been studied adequately.

ORIGINS OF PRIVATIZED AGRICULTURAL EXTENSION SERVICES WORLDWIDE

During the 1980s, there were three distinct developments in agricultural extension: private extension, commercialization, and privatization of extension services. However, the existence of private extension may be traced long before 1980s. Due to the emergence of commercialized farming in developed countries, the private character of technologies and patent regimes made the private sector visible in extension services. This private extension sector was dominated by private for-profit entities such as agribusiness firms, private consultants, and the media. The private nonprofit sector was dominated by the NGOs and farmers' associations. The commercialization of extension services has occurred in developed and as well as developing countries (LeGouis, 1991). Under commercialization, as opposed to privatization, the agency remains public but specific services are sold for a fee, while other services may remain public goods (Dinar, 1996). Two sets of issues have confronted the development of privatization and private-sector provision of extension. We briefly review them below.

Privatization and commercialization of extension

During the 1970s to 1990s, there were many attempts to privatize and/or commercialize extension services in many developed countries, especially European countries and a few Latin American countries. These developments also influenced many other countries to debate their national extension systems and the possibility of including elements of private extension. As part of the debate, there were a few studies on attitudes and opinions of farmers regarding privatization of extension services and their willingness to pay for extension services. Baxter (1987) opined that the private extension agencies do not usually apply their resources to the food crops that are fundamental to farmers, but to those who produce cash crops, and hence the developing countries were not ready for any substantial withdrawal of public support to extension. A commonly encountered weakness of private-sector input supply agencies involved in transfer of technology is that little attention was given to low-input, sustainable agricultural technologies, including environmental resource conservation technologies. In addition, public extension personnel are engaged not only in the task of transfer of information to their clients, but also in fostering the empowerment of farming community by organizing, motivating, and guiding farmers' groups through group management approaches (Harter and Hass, 1992). These types of extension approaches may not find favor with private consultancy firms, as they reduce the number of "days sold per consultant each year." Driven by the interests of those clients who are able to pay the bills, they may no longer be agencies responsive to the public interest as a whole.

Sulaiman and Gadewar (1994) expressed concerns that the education role played by public extension agencies at present may be lost if the extension services are privatized. Privatization is likely to widen the already prevailing socioeconomic inequality in different parts of the country. In general, the ability of farmers to pay for extension services is very weak, because 70% of the net sown area is rainfed, and farmers are small, marginal, and resource poor. Extension is a very serious business that cannot be left to commercial agencies. Public extension needs to address issues of sustainability, the environment, and equity. Van den Ban and Hawkins (1996) stated that privatization of government extension services has advantages as well as disadvantages. Local situations will determine whether the advantages outweigh the disadvantages. Advice from a privatized system may be more effective because the farmer can select an advisor who is best able to help. Further, farmers may be more inclined to follow advice for which they have paid. Privatization of extension may hamper the free flow of information, as extension agents in a privatized extension service are inclined to concentrate on larger farmers who can afford to pay their fees.

According to Rivera and Cary (1997), privatization may have some attendant disadvantages because of unequal access to resources and because of the diversity of agencies and the associated difficulty of coordinating external groups and government departments. Private delivery agents may be less responsive to government policy direction, and there may be problems around linkage with public applied research organizations.

In general, experts believe that a more commercialized approach broadens the focus of extension personnel and makes an extension service more responsive to client needs and changing economic and social conditions. But other immediate implications of privatization appear to include a tendency toward a reduction in linkages among organizations and farmers in the exchange of agricultural and other relevant information, a tendency to enhance large-scale farming, a diminishing emphasis on public good information and the advancement of knowledge as a saleable commodity, and a trend toward agricultural development services that cater primarily to large-scale farming.

Attitude toward private extension/privatization

How farmers perceive the provision of private extension has been a major determinant of the expansion of private extension. There are a number of research studies that have been conducted to find out the attitudes of farmers toward the privatization of agricultural extension services in India (Saravanan, 1999; Venkata Kumar et al., 2000; Hanchinal et al., 2001; Jiyawan et al., 2009; Jasvinder et al., 2014). Results revealed that a substantial number of farmers had a favorable attitude toward privatization. Similar studies have also been conducted in Iran (Farokhi, 2002) and Sri Lanka (Rohana, 2005).

In a study conducted among 720 farmers from three states of India, Sulaiman and Sadamate (2000) found that about 48% of farmers expressed a willingness to pay for agricultural information. Saravanan and Veerabhadraiah (2003), in Karnataka (India), reported that 36.7% of the clientele of Farmers Communication Centers were willing to pay for the services provided by a public organization, which ranged from Rs. 50 to Rs. 300. The main areas of information for which farmers were willing to pay were cultivation practices for fruit crops, marketing information, and post-harvest technology. In Zimbabwe, Foti et al. (2007) found that only 4.6% of the farmers were willing to pay for extension services, and 95.4% of the farmers were not willing to pay for extension.

Ali et al. (2008) in Iran reported that only 24.7% of farmers were willing to pay consulting engineers for wheat, and 75.3% of the farmers were not willing to pay. Oladele (2008) in Nigeria reported that the most prominent services farmers were willing to pay for were providing information to women farmers (34%), identifying rural problems (38%), training VEA (33%), supervising women's activities (43%), arranging input supply (36%), processing loans (32%), organizing group meetings (38%), giving advice on agricultural problems (33%), teaching home management for children and nutrition (29%), organizing farmers' seminars, group discussions (26%), and liaison with farm machinery sources (34%).

Budak et al. (2010) in Turkey concluded that 52.5% of livestock producers were willing to pay for agricultural extension services. Francis et al. (2010) indicated that in Uganda 35% out of 5,363 farmers and 40% out of 3,318 farmers were willing to pay for extension services related to crops and animal husbandry, respectively. The willingness to pay for extension services was slightly higher among animal husbandry farmers than the crop farmers. Singh et al. (2011) in central Uttar Pradesh (India) stated that 76% of farmers were ready to pay for advice on plant protection measures, followed by 63% of farmers who were willing to pay for advice on weed management, and 60% agreeing to pay for animal husbandry management.

Makdisi Fadi and Marggeaf Rainer (2011) in Germany found that 82.3% of respondents were willing to pay for certified farm animal products, while the rest (17.7%) objected to paying more. Ozor et al. (2011) in Nigeria revealed that the vast majority (95.1%) of farmers were willing to pay for improved extension services; only 4.9% indicated unwillingness to pay. Ulimwengu and Sanyal (2011) reported that in

Uganda 47.1% farmers were willing to pay for soil fertility management, 45.9% were willing to pay for improved produce quality/varieties, 43.3% were willing to pay for disease control, and 43.2% were willing to pay for crop protection.

Thus there is a large body of evidence to show that farmers are willing to pay for information services. However, the sustainability of such payment for services may depend on the quality and reliability of the service.

DOCUMENTATION OF CASE STUDIES ON PRIVATIZATION AND NATIONAL REFORM PROCESS

There is a large body of literature on agricultural extension and advisory services and their benefits to farmers. However, a smaller body has focused on analysis of the structure of extension provision in its various forms. Rivera and Alex (2004) compiled 12 case studies from Chile, Ecuador, Estonia, Germany, Honduras, Mali, Niger, Pakistan, South Africa, Venezuela, Uganda, and the United Kingdom where privatization has been a feature of the national or regional reform process over the past 10–15 years. From these case studies, it was realized that privatization of rural services involves the development of new partnerships and associated capacities between government agencies and nongovernmental and private-sector actors (Connolly, 2004), as opposed to the ceding of total or substantial ownership and operational control from the government to the private sector. Table 3.1 highlights key earlier publications critical to a comprehensive understanding of the transition from public extension to alternative forms of extension provision, as they were the first to explore this area. The literature

Table 3.1 Landmark publications on alternative and private extension

Sl. No.	Authors and year	Landmark publications and articles
1.	Roling (1982)	Alternative approaches in extension. In: Jones, G.E., Rolls, M.J., 1982. Progress in Rural Extension and Community Development, vol. I. John Wiley & Sons, Ltd.
2.	Rivera and Gustafson (1991) (Eds.)	Agricultural Extension: World Wide Institutional Evolution and Forces for Change. Elsevier Publishers, Amsterdam, the Netherlands.
3.	Harter and Hass (1992)	Commercialization of British extension system: Promise or primrose. *Journal of Extension Systems* 8, 37–44.
4.	Ingram (1992)	The United Kingdom experience in the privatization of extension agricultural delivery systems: Public and private roles in agricultural development. Proceedings of the Twelfth Agricultural Sector Symposium, Jock R. Anderson and Cornelis de Haan (Eds.). World Bank Publications, Washington, DC.

(Continued)

Table 3.1 (Continued)

Sl. No.	Authors and year	Landmark publications and articles
5.	Bloome (1993)	Privitization lessons for U.S. extension from New Zealand and Tasmania. *Journal of Extension* 31, 24–6.
6.	Umali and Schwartz (1994)	Public and private agricultural extension: Beyond traditional frontiers. World Bank Discussion Paper No. 236, World Bank publications, Washington, DC.
7.	Ameur (1994)	Agricultural extension: A step beyond next step. World Bank Technical Paper No. 247, Washington, DC.
8.	Sulaiman and Gadewar (1994)	Privatizing farm extension: Some issues. International Workshop on "Alternative and cost effective approaches for sustainable agriculture: Methodological issues." Proceedings and Selected Theme Papers, Organized by Ford Foundation, FAO and TNAU, Coimbatore, India. September 14–17, 1994, 55–60.
9.	Dinar (1996)	Extension commercialization: How much to charge for extension services? *American Journal of Economics* 78 (1), 1–12
10.	Keynan et al. (1997)	Co-financed public extension in Nicaragua. *World Bank Research Observer*, 12 (2).
11.	Umali-Deininger (1997)	Public and private extension: Partners or rivals? *World Bank Research Observer. International Bank for Reconstruction and Development, World Bank* 12(2), 230–224.
12.	Carney (1998)	Changing Public and Private Roles in Agricultural Service Provision. Published by the Overseas Development Institute, London.
13.	Saravanan (1999)	A Study on Privatization of Agricultural Extension Services. MSc thesis, Department of Agricultural Extension, University of Agricultural Sciences (UAS), GKVK, Bangalore, India.
14.	Sulaiman and Sadamate (2000)	Privatizing agricultural extension in India. Policy Paper 10, National Centre for Agricultural Economics and Policy Research (NCAP), New Delhi, India.
15.	Van den Ban (2000)	Different ways of financing extension. ODI, Agricultural research and extension network paper, London.
16.	Chandrashekara (2001) (Ed.)	Private Extension: Indian Experiences. National Institute of Agricultural Extension Management (MANAGE), Hyderabad, India.
17.	Rivera and Alex (2004)	Extension system reform and the challenges ahead. *The Journal of Agricultural Education and Extension*, 10 (1), 23–36.
18.	Rivera and Alex (2004)	Privatization of Extension Systems: Case Studies of International Initiatives (Volume 2). Agriculture and Rural Development Discussion Paper Number 9, Extension Reform for Rural Development.

Source: Authors' compilation.

cited in Table 3.1 is a good foundation for understanding the need for and benefits of pluralistic extension systems.

During the 2000s, many national governments in the developing countries realized that public extension alone may not be able to cater to the needs of the farming community and hence encouraged public–private partnerships in the agricultural extension. The importance of complementary and/or supplementary role of private sector players in extension was emphasized, including agribusiness firms, private consultancies, media, and nongovernmental organizations. Hence, pluralistic extension systems were given considerable importance in agricultural extension policy documents and programs.

Case 1: Benin (Agbo et al., 2008)
Commercialization of Extension Services (Fee for Extension)

Agricultural extension services are mainly freely offered to producers within the framework of the diffusion of the majority of technologies. For some organized sectors such as cotton, producers pay for the extension services in particular CeRPA through their association, which has a contract with that structure. On each kilogram of grain cotton sold, 10–20 Fcfa is taken. This share from the producers enters a unit called funds of critical functions (Fcfa) used to finance extension, repairing of rural tracks, funding research on cotton, etc. Thus the producer indirectly participates in the financing of his supervision.

Private-Sector Extension

Today the private sector is more particularly responsible for:

- *Supplying input and other service provisions*: The CARDER indeed stopped their activities of providing inputs to the farmers; in the cotton zone, these were taken over by the private sector.
- *Transformation and marketing*: The transfer of the commercial activities of CARDER to the private sector (entrepreneurs, farmers organization, NGO) was mostly carried out.
- *Support services and in particular the agricultural support service*: The involvement of the private sector is slow and progressive, particularly in the sectors of the agricultural advice and training, services for which professional organizations should ensure the responsibility by themselves or by private providers.

Other actors from the private sector also have fostered the functions: companies in the agro-food sector and the agricultural production support sector as well as NGO, partly taking over functions in which the state was disengaged or for which the state and the MAEP could not mobilize specific competencies because they are currently out of their mandate.

Extension by NGOs

Although the agricultural sector is liberalized, one does not yet observe in Benin experiments involving private extension service at the request of the producers. Nevertheless, the majority of NGOs active in the agricultural extension service are actors in development projects/

programs, who themselves develop their strategy for extension in the field. This is the current case with conservation and management of the Natural Resources Program of Benin-German cooperation, which in its intervention zone in the northeast of the country uses an NGO as service providers for the supervision of its beneficiary groups in the villages.

Agbo, B.P., 2008. Agoundote desire and midingoyi gnonna Soul-Kifouly, 2008, Benin. In: Saravanan, R. (Ed.), Agricultural Extension: Worldwide Innovations. New India Publishing Agency (NIPA), New Delhi, India.

Case 2: Côte d'Ivoire (Alphonse, 2008)
Private Extension

In Côte d'Ivoire, extension by the private sector was organized around poultry, rubber, palm oil, sugar, and cotton. It relates to a given sector and is not only to provide agricultural advisory services, but also to distribute key agricultural inputs such as fertilizers, pesticides, etc. Further, rubber growing companies (SAPH, SOGB), palm oil (PALMCI, PALMAFRIQUE, PHCI, etc.), cotton (CIDT, LCCI, IVOPIRE COTTON), and others (FACI, SIPRA, IVOGRAINS) assist farmers engaged in these crops.

Private consultancy services are also available because of peasant organizations that hire experts to adopt advanced technologies. Members pay a membership fee. With that amount the farmers' organization hires expert advice and provide funds for technical guidance and also for the farm visits of the extension advisors. This is prevalent mostly in rubber, oil palm, coffee, cocoa, and rice crops. Likewise, farm equipment and agricultural input providers are very much interested in ensuring the satisfaction of their customers, investing much effort and time in providing advice on the use of their farm inputs. In terms of private extension, the costs of the services carried out by the extension companies are taken into account in pricing of farm produce.

Public–Private Partnership

The agricultural sector is very large and diverse in nature; ANADER alone does not have the means necessary for extension service to be provided to all producers. The state is required to rely on the private sector and NGOs to take care of producers in all their aspects. For well-organized, industry-based crops such as palm oil, rubber, cotton, etc., the state conceded the entire responsibility of extension services to companies that exploit these crops. For less structured crop sectors such as coffee–cocoa crops, etc., the state takes care of a part of responsibility and a part of the extension service is provided by the private firms and NGOs with financing by development funds (FDFP, FIRCA, or international donors).

Alphonse, K.K., 2008. Côte d'Ivoire. In: Saravanan, R. (Ed.), Agricultural Extension: Worldwide Innovations. New India Publishing Agency (NIPA), New Delhi, India.

Case 3: Ghana
Private-Sector Extension

Private for-profit sector (input, suppliers, and agribusiness firms) involvement in agricultural extension service provision is increasing in Ghana. Private-sector companies provide information and advice (and sometimes technical advice by field visit) on the product as an unidentified component of sales price (McNamara et al., 2013).

Olam International

Through a contract farming arrangement, it provides advisory services to farmers and conducts training on food and agricultural practices on 100 demonstration plots in cotton cultivation by hiring 100 extension agents (McNamara et al., 2013).

Farmer organization

Kuapakokoo cocoa production union has 65,000 members. Its own marketing/trading business unit was formed in 1980. The union operates in 57 districts of the five southern regions of Ghana. It has 32 extension personnel for advisory services. The union also conducts some technical and social research and also undertakes capacity building and infrastructure development.

NGOs

A large number of NGOs depend on donor funding for projects and activities. Most of the NGOs work with the local champions/lead farmers (nucleus farmers)/promoters who work with a group of farmers to promote new techniques/crops. The lead farmer may work with existing groups/FBOs/or create new groups/FBOs. Most of the NGOs use demonstrations, training, field days, and encourage farmer-to-farmer communication. In some cases, videos are used. Radio is most popular among farmers and some NGOs. NGOs are using local FM stations to create awareness. Virtually all NGOs seek to engage the district-level MoFA Agriculture Extension Agents (AEAs) in their projects.

Alphonse, K.K., 2008. Côte d'Ivoire. In: Saravanan, R. (Ed.), Agricultural Extension: Worldwide Innovations. New India Publishing Agency (NIPA), New Delhi, India.

Case 4: India (Saravanan, 2008)

In India, agricultural consultants/consultancies, agribusiness firms, input dealers, mass media (radio, TV, and ICTs), and farmers' associations are engaged in providing single or integrated services to farmers. These may deal with information, input supply, infrastructure arrangement, and marketing services (Chandrashekara, 2001).

Agricultural Consultancies

Saravanan (1999) reported that agricultural consultants and consultancy agencies emerged after the mid-1990s in Coimbatore district of Tamil Nadu state of India. Most of the consultancies are nonregistered, mostly run by a single technical person, covering a small area, mainly concentrating on all aspects of horticulture crops. Mostly nonagricultural sector people and a few big farmers are the main clients of private consultancies. Similar types of agri-clinics to provide testing facilities, diagnostic and control services, and other consultancies on a fee-for-service basis have been set up by trained agricultural graduates in large numbers since 2003. For this purpose, agricultural graduates were trained through agri-clinics and agribusiness centers, a scheme of the central government. These trained individuals were supported to take up individual and group ventures, which were financed by bank loans with a 25% central government subsidy. Through the agri-clinics and agribusiness centers scheme, large numbers of agricultural graduates were trained, and 5,008 success stories were reported up to 2007 (www.agriclinics.net).

Similarly, extension activities of the several agribusiness firms, farmers' associations, and cooperatives and NGOs were documented.

Saravanan, R. (Ed.), 2008. Agricultural Extension: Worldwide Innovations. New India Publishing Agency (NIPA), New Delhi, India.

CHALLENGES AND ISSUES FACING THE DEVELOPMENT AND EXPANSION OF PRIVATE EXTENSION

The development and expansion of private extension continue to face several challenges. These are highlighted below, as these issues are often identified in the detailed case studies presented in later chapters of this book.

Non-Conducive Policy Environment: In many developing countries, there is limited policy support and incentive to carry out extension activities by the private sector.

Narrow Focus of Private Extension: Most private extension service providers concentrate on only a few commercial crops. Mostly large landholding and commercial farmers were benefited.

Limited Capacity of Private Extension: Private extension operates with the minimum of human resources for agricultural extension and advisory services. Hence, it is difficult to reach large number of farmers.

Farmers' Mindset: In most developing countries small and marginal landholding farmers view extension as a welfare activity of the state, and thus they are unwilling to share the responsibility for receiving the services.

Public–Private Partnership: The institutional culture of the public and private sectors offers limited scope to develop and sustain partnership over a longer period.

Public Good Nature of Extension Services: The public good nature of the majority of extension services hinders the emergence of the private sector.

Limited Research Capacity of Private Extension: The private sector has research expertness in only a few selected areas, which hinders the capacity of private extension to engage the entire agriculture value chain and advisory services provision.

Food Security and Natural Resource Management: Private extension generally has limited interest in investing in field crops and natural resource management issues because they may not get much profit.

Subsistence and Uncertain Nature of Agriculture: Small and marginal landholding farmers in many developing countries still practice subsistence farming along with unpredictable weather and market scenarios, which gives them limited income. As a result, farmers are unable to invest in private extension.

CONCLUDING REMARKS

In this chapter we reviewed historical trends in the development of private extension as a basis for discussion of the case studies presented in the ensuing chapters. Private extension approaches begin with a broader intention of the policymakers to let the private sector operate side by side with the public extension, for the most part in areas where the public sector is unable to provide quality services. The rest of the chapters in this book provide detailed accounts of several recent case studies that analyze issues and challenges facing the development of private extension systems.

REFERENCES

Ali, S., Ahmad, M., Ali, T., Islam-ud-Din, Iqbal, Z.M., 2008. Farmers' willingness to pay (WTP) for advisory services by private sector extension: the case of Punjab. Pak. J. Agric. Sci. 45 (3), 107–111.
Ameur, C., 1994. Agricultural extension: a step beyond next step. World Bank Technical Paper No. 247, Washington, DC.
Antholt, C.H., 1994. Getting ready for the twenty-first century: Technical change and institutional modernization in agriculture. World Bank Technical Paper, No. 217, World Bank Publication, Washington, DC.
Baxter, M., 1987. Emerging priorities for developing countries in agricultural extension. In: Rivera, W.R., Schram, S.G. (Eds.), Agricultural Extension Worldwide: Issues, Practices and Emerging Priorities. Croom Helm, New York.
Berdegué, J.A., 1997. Organisation of Agricultutal Extension and Advisory Services for Small Farmers in Selected Latin American Countries. Paper Presented at the Development Workers Course: Technology development and transfer: How to Maximise the Influence of the User. Can Alternatives to the Training and Visit system be found? Landbrugets Center for Efteruddannelse, 7–11 April 1997, Greve, Denmark.
Bloome, P.D., 1993. Privitization lessons for U.S. extension from New Zealand and Tasmania. J. Ext. 31, 24–26.
Budak, B.D., Budak, F., Kacira, O.O., 2010. 'Livestock producers' needs and willingness to pay for extension services in Adana Province of Turkey. Afr. J. Agric. Res. 5 (11), 1187–1190.
Carney, D., 1998. Changing Public and Private Roles in Agricultural Service Provision. Published in the Overseas Development Institute, London.
Chandrashekara, P. (Ed.), 2001. Private Extension: Indian Experiences National Institute of Agricultural Extension Management (MANAGE), Hyderabad, India, p. 17.
Connolly, M., 2004. Private extension and public–private partnerships: privatized, contracted and commercialized approaches In: Rivera, W. Alex, G. (Eds.), Privatization of Extension Systems: Case Studies of International Initiatives, vol. 2. Agriculture and Rural Development Discussion Paper Number 9, Extension Reform for Rural Development.
Dinar, A., 1996. Extension commercialization: how much to charge for extension services. Am. J. Agric. Econ. 78 (1), 1–12.

Farokhi, S., 2002. Attitudes of experts and farmers of Ilam province to privatization of agricultural extension. M.Sc. Thesis of Tarbiat Modares.

Foti, R., Nyakudya, I., Mayo, M., Chikuvire, J., Mlambo, N., 2007. Determinants of farmers demand for "fee-for-service" extension in Zimbabwe: the case of Mashonaland Central province. J. Int. Agric. Ext. Educ. 14 (1), 95–104.

Francis, M., Ronald, M.F., Geofrey, O., 2010. Willingness to pay for extension services in Uganda among farmers involved in crop and animal husbandry. Contributed paper presented at the Joint 3rd African Association of Agricultural Economists (AAAE) and 48th Agricultural Economists Association of South Africa (AEASA) Conference. Cape Town, South Africa.

Gustafson, D.J., 1991. The challenge of connecting priorities to performance: One state's response to the forces for change in U.S. extension. In: Riviera, W.M., Gustafson, D.J. (Eds.), Agricultural Extension: Worldwide Institutional Evolution and Forces for Change. Elsevier Science Publications, Amsterdam.

Hanchinal, S.N., Sundaraswamy, B., Ansari, M.R., 2001. Attitude and preferences of farmers towards privatisation of extension service. In: Chandra Shekara, P. (Ed.), Private Extension in India: Myths, Realities, Apprehensions and Approaches. National Institute of Agricultural Extension Management (MANAGE), Hyderabad.

Harter, D., Hass, G., 1992. Commercialization of British extension system: promise or primrose. J. Ext. Syst. 8, 37–44.

Howell, J., 1985. Recurrent Costs and Agricultural Development. Overseas Development Institute, London.

Jasvinder, K., Shehrawat, P.S., Peer, Q.J.A., 2014. Attitude of Farmers Towards Privatization of Agricultural Extension Services. Department of Extension Education, CCS Haryana Agricultural University, Hisar, India.

Jiyawan, R., Jirli, B., Singh, M., 2009. Farmers' view on privatization of agricultural extension services. Indian Res. J. Ext. Edu. 9 (3).

Keynan, G., Olin, M., Dinar, A., 1997. Co-finaced public extension in Nicaragua. World Bank Res. Obs. 12 (2).

LeGouis, M., 1991. Alternative financing of agricultural extension: recent trends and implications for the future. In: Riviera, W.M., Gustafson, D.J. (Eds.), Agricultural Extension: Worldwide Institutional Evolution and Forces for Change. Elsevier Science Publications, Amsterdam.

Makdisi, F., Marggeaf, R., 2011. Consumer Willingness-to-Pay for Farm Animal Welfare in Germany—The Case of Broiled. Gewisol, Department of Agricultural Economics and Rural Development, University of Gottingen, Germany.

Matanmi, B.M., Ladele, A.A., 1996. Participation of private organizations in agricultural development: Lessons from the extension type activities of the Alimontos Congelados Monte Bellos, S.A. (ALCOSA) in the Guatemala and the Shell Petroleum in Nigeria. J. Rural Dev. Admin. 28 (2), 39–50.

McNamara, P., Dale, J., Keane, J., Ferguson, O., 2013. Strengthening Pluralistic Agricultural Extension Ghana, Report on the MEAS Rapid Scoping Mission. Published by the Modernizing Extension and Advisory Services (MEAS).

Oladele, O.I., 2008. Factors determining farmers willingness to pay for extension services in Oyo State', Nigeria. Agricultura Tropica Et Subtropica 41 (4), 165–171.

Ozor, N., Garforth, J.C., Madukwe, C.M., 2011. Farmers' willingness to pay for agricultural extension service: evidence from Nigeria. J. Int. Dev. Available from: http://dx.doi.org/10.1002/jid.

Picciotto, R., Anderson, J.R., 1997. Reconsidering agricultural extension. World Bank Res. Obs. 12 (2), 249–259.

Rivera, W.M., 2001. The invisible frontier: The current limits of decentralization and privatization in developing countries. In: Brewer, F. (Ed.), Agricultural extension: An international perspective. Erudition Press, Rome.

Rivera, W., Cary, J., 1997. Privatizing agricultural extension. In: Swanson, B.E., Bentz, R.P., Sofranko, A.J. (Eds.), Agricultural extension: a reference manual Food and Agriculture Organization of the United Nations, Rome.

Rivera, W.M., Alex, G., 2004. Privatization of Extension Systems: Case Studies of International Initiatives. Discussion Paper 9. World Bank Publications, Washington, DC.

Rivera, W.M., Gustafson, D.J. (Eds.), 1991. Agricultural Extension: World Wide Institutional Evolution and Forces for Change. Elsevier Publishers, Amsterdam, the Netherlands.

Rohana, P.M., 2005. Attitude of Agricultural Scientists Extension Personnel and Future Towards Commercialization. Faculty of Agricultural Sciences, Sabargamuwa University, SriLanka.

Roling, N., 1982. Alternative approaches in extension. In: Jones, G.E., Rolls, M.J. (Eds.), Progress in Rural Extension and Community Development, vol. 1. John Wiley & Sons, Ltd.

Saravanan, R., 1999. A Study on Privatization of Agricultural Extension Services, MSc thesis. Department of Agricultural Extension, University of Agricultural Sciences (UAS), GKVK, Bangalore, India.

Saravanan, R., Shivalinge Gowda, N.S., 1999. Development of a scale to measure attitude towards privatization of agricultural extension service. Trop. Agric. Res. 11, 190–198.

Saravanan, R., Veerabhadraiah, V., 2003. Clientele Satisfaction and Their Willingness to Pay for Public and Private Extension Services, Tropical Agricultural Research, vol. 15. PGIA, Peradeniya University, SriLanka.

Schwartz, L., 1994. The role of the private sector in agricultural extension. Economic analysis and case studies. Agricultural Research and Extension Network, Paper No. 48, Overseas Development Institute, London.

Singh, A.K., Narain, S., Chauhan, J., 2011. Capacity of farmers to pay for extension services. Indian Res. J. Ext. Educ. 11 (3), 60–62.

Sulaiman, R.V., Gadewar, A.V., 1994. Privatizing farm extension—some issues. International Workshop on Alternative and Cost Effective Approaches for Sustainable Agriculture: Methodological Issues—Proceedings and Selected Theme Papers. Organised by Ford Foundation, FAO and TNAU, Coimbatore, India. September 14–17, pp. 55–60.

Sulaiman, V.R., Sadamate, V.V., 2000. Privatising Agricultural Extension in India, Policy Paper 10. National Center for Agricultural Economics and Policy Research (NCAP), New Delhi, India.

Ulimwengu, J., Sanyal, P., 2011. Joint Estimation of Farmers' Stated Willingness to Pay for Agricultural Services. IFPRI Discussion Paper 01070. International Food Policy Research Institute, West and Central Africa.

Umali, D.L., Schwartz, L., 1994. Public and Private Agricultural Extension: Beyond Traditional Frontiers. World Bank Publications, Washington, DC.

Umali-Deininger, D., 1997. Public and private agricultural extension: partners or rivals? World Bank Res. Obs. 12 (2), 203–224.

Van den Ban, A.W., 2000. Different ways of financing extension, ODI, Agricultural Research and Extension Network Paper, London.

Van Den Ban, A.W., Hawkins, H.S., 1996. Agricultural Extension. Blackwell Science Ltd. Pub., Oxford, pp. 256–258.

Venkatta Kumar, R., et al., 2000. Privatization of agricultural extension system in India presented at National seminar on private extension. July 28–29, 2000, MANAGE, Hyderabad (AP), India.

Wilson, M., 1991. Reducing the costs of public extension services: Initiatives in Latin America. In: Riviera, W.M., Gustafson, D.J. (Eds.), Agricultural Extension: Worldwide Institutional Evolution and Forces for Change. Elsevier Science Publications, Amsterdam.

CHAPTER 4

Private Sector Extension with Input Supply and Output Aggregation: Case of Sugarcane Production System with EID Parry in India

Suresh Chandra Babu

International Food Policy Research Institute (IFPRI), Washington, DC, USA

Contents

INTRODUCTION

Privatizing extension services through government reforms has yielded mixed results in developing countries. At the same time, in an effort to connect smallholder farmers to markets, various forms of extension services have emerged in developing countries (Birner and Anderson, 2007; Babu et al., 2013). While the extension provided by

Knowledge Driven Development.
DOI: http://dx.doi.org/10.1016/B978-0-12-802231-3.00004-8

private companies will not fully address the needs of the smallholder farmers, it could form a sustainable way to wean the farmers off the inefficient public extension services. Further, private extension has emerged on its own accord in many developing countries as they try to enhance the commercial crop sector. While research resources are invested in developing new technologies for farmers, realization of the benefits of these technologies will require innovations in extension approaches. Private companies that rely on the crop outputs of the farmers for their processing and marketing operations are intrinsically motivated to provide extension and other services to farmers. For them, one way to develop trust with the farmers who supply their raw materials is to provide them with advice on production techniques (Anderson and Feder, 2004; Anderson et al., 2006).

In this case study we try to address the following questions with the objective of understanding institutional innovations in extension delivery by private entities that work with farmers: What are the private sector's motivations to provide extension services beyond the regular crop purchasing contracts as in typical contract farming arrangements? How desirable is private extension in the presence of a public extension system that serves the farmers in the same production system? What are the similarities and complementarities between the public and private extension systems? What opportunities are there for public–private–people partnerships in extension provision? What policy changes trigger innovation in farm-level extension? How does a private company provide extension service? Who do they engage with to provide extension services? Who absorbs the costs of the extension services? What are the specific services provided and at what stage of production are they offered?

We address these questions in the context of the evaluation guidelines developed for the purpose of this study, which focus analysis on the relevance, effectiveness, efficiency, impact, and sustainability of extension services.

We also ask the following questions: What are the implications and impact of private extension on the productivity, income, livelihood, and poverty levels of the farmers served? What are the perceptions of the government, public extension workers, policymakers, and the participating farmers of the private system of extension? How sustainable, replicable, and scalable are the systems of extension undertaken by the private sector?

In order to develop this case study, field visits were conducted during the month of March 2013 in the Cuddalore region of Tamil Nadu, India. A set of guided discussions and interviews were conducted with farmers, farmer entrepreneurs, the company's extension staff, public extension staff, program leaders of government programs, input dealers, researchers at the public research station, and community leaders.

The case study is organized as follows. The next section briefly introduces the sugarcane production system in India and its extension needs and contractual arrangements with the sugar mill industry. The method of extension used by the private company is described in the section "Description of the Extension Approach." The section "Analysis of the Private Extension Services with Respect to Relevance, Efficiency, Effectiveness, Equity, Sustainability, and Impact" assesses the relevance, effectiveness, efficiency, sustainability, and impact of the private extension. The perspectives of various actors and the community are given in the section "Comparison with Public Extension in India." Lessons learned are given under "Lessons for Replication and Scaling Up." Concluding remarks form the last section.

In this case study we take an in-depth look at EID Parry India, one of the earliest private sugar mills in the country, to see how they innovate to provide extension and advisory services to their contracted farmers. Sugarcane production in India is one of the country's key commercial agricultural enterprises, with direct implications on smallholder income and livelihoods. India cultivates 5 million hectares of sugarcane and produces 342 million tons of sugarcane. The highest producing states are Uttar Pradesh, Maharashtra, and Tamil Nadu. Most of the sugarcane production in India is used to produce white sugar. Until recently, government regulations forced farmers to send their sugarcane to their local area sugar mill. This supported the sugar mills by ensuring a regular supply of sugarcane. This regulation also meant that farmers faced a monopsony buyer for their cane.

DESCRIPTION OF THE VALUE CHAIN (PRODUCTION AND MARKETING SYSTEM)

The sugarcane value chain in India revolves around the sugar companies and their operations. As far as the sugarcane farmer is concerned, the factory is the final consumer of the cane. Farmers are paid for their sugarcane directly by the sugar company. There are currently a variety of sugar companies in operation within any given area. These include cooperative sugar mills where farmers can be members, government-run sugar mills, and privately owned sugar mills. Sugar companies supply sugar to wholesale markets and are also required to supply a certain amount to the government for the public distribution system. Various actors and players in the sugarcane value chain are briefly described below (Figure 4.1).

Sugarcane company

EID Parry produces white sugar by crushing sugarcane; the sugar is used by a subsidiary company, Parry Sweets, a confectionary producer and marketer. The confectionary industry requires a constant supply of sugar, and sugar mills, in turn, require a constant supply of sugarcane from the farmers. EID Parry has five sugar producing plants in various parts of Tamil Nadu. This case study is based on information gathered from one of these plants, the Nellikuppam Sugar plant, which is one of the oldest in the country, having been started during the British rule of India. The company enters

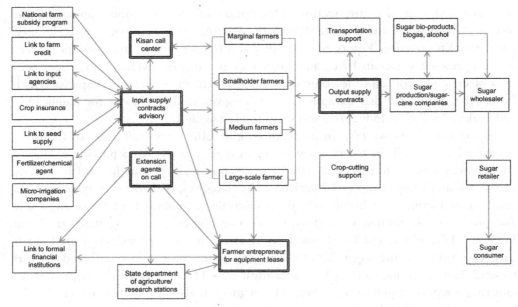

Figure 4.1 Sugarcane value chain: farmer to sugarcane company. *Source: Author's compilation.*

into a contract with a farmer who is willing to grow sugarcane for them. The contract specifies what the farmer will get in the form of support to grow sugarcane and how the produce will be purchased. The price is dependent on the quality of the cane, the range of quality levels being specified in the contract.

For optimal profit margins, the private company needs a regular supply of sugarcane grown by the farmers in adequate quantity to keep the factory running for 300 days in a year. It needs high-quality sugarcane on a consistent basis, which requires updating farmers' knowledge through education of the farmers. In order to keep the farmers in sugarcane production, the company has increased the quality and coverage of its extension services, moving away from being only the purchaser to providing extension and related services to the farming community. Although the company is a monopsony purchaser of sugarcane, the farmers have the option to grow other crops. Thus the company must pay a fair price to the farmers to retain their interest in sugarcane farming. Although the sugarcane value chain in India revolves around the crushing factories, there are other key players that make the value chain function. They are described briefly below.

Sugarcane farmers

Sugarcane is grown by farmers either as a main crop or as a cash crop within a diversified portfolio. For smallholders in the study region, those who own 2–5 acres, sugarcane continues to be the main crop. They are characterized by continuous grow-ing of sugarcane with and without ratooning. For these farmers, the extension services provided by the sugarcane company are crucial to keep sugarcane farming a viable activity. Some farmers in this category have moved away from sugarcane farming in the last 20 years due to increased opportunities in nonfarm activities and also partly due to younger generations moving out of agriculture. Many farmers have moved to growing rice and other crops to meet the food security needs of their family. These trends are a concern for the sugarcane company as they reduce the supply of raw cane, reduce the company's economies of scale, and make it difficult for the company to reach its goal of keeping the factory running for at least 300 days in a year to obtaining optimal productivity.

Medium farmers, those who own 5–10 acres of land, have a little more flexibility to grow rice and groundnuts in addition to sugarcane; to some extent they are in a bet-ter position to keep a part of their land under casuarina—a long-term fuel-wood tree which gets a good price in the market. These farmers are able to purchase sugarcane inputs from sources other than the private company. Their primary concern is the timely availability of extension services from the public sector, which is currently pro-vided by the private sector. Increasing the share of sugarcane in these farmers' crop portfolio is the sugarcane company's goal. Farmers who own more than 10 acres of land have more flexibility to allocate their land to sugarcane and other crops. They are also in a better position financially to invest in drip irrigation. They tend to reap a higher yield of sugarcane, which is also of better quality than smallholder farmers' cane. The larger-scale farmers are preferred by the processing company because of the trans-action costs associated with each additional contract.

Micro-irrigation companies

Several micro-irrigation companies have taken advantage of the subsidy provided to sugarcane farmers to supply them with equipment and technical advice, making sugar-cane farming a viable and cost-effective undertaking, particularly in water-deficient regions. The micro-irrigation companies are aided by the public extension system in identifying potential farmers. EID Parry supports this process by aiding the farmers to make their part of the investment. Converting from flood irrigation to drip irrigation also helps farmers to conserve fertilizer since the fertilizer can be applied through the drip irrigation system.

Mechanization entrepreneurs

In order to solve the labor shortage, entrepreneurial farmers are identified and given loans to purchase land preparation equipment and harvesters, which helps farmers to complete their work on time. These entrepreneurs engage in contracts with farmers through the sugarcane company, which pays for their services when the sugarcane is harvested and supplied to the company.

Research system and the seedling multipliers

A sugarcane research station located in the region plays a critical role in growing seedlings of the latest varieties to sell to the farmers. They also act as knowledge centers, advising farmers on soil testing services. The sugarcane company uses the new technologies developed by the research center to enhance varieties of cane after adaptive research trials on its own farm.

Credit institutions

Formal financial institutions collaborate with the sugarcane company to finance farmers with short-term credit. Repayment of the loans is made by the company when the sugarcane is sold. This partnership has removed the need for local money lenders in these villages, at least for consumption purposes.

Input dealers

Input dealers supply fertilizers and other chemicals to farmers on a cash and credit basis. Again, repayment occurs when the sugarcane is purchased by the company via a direct agreement with the input dealer. The company also has its own shops to sell its fertilizers brands to the farmers.

Public extension system

The public extension system engages with sugarcane farmers primarily to distribute the subsidies for small equipment and conduct other government intervention programs. In some ways, the public extension agents serve as a watchdog, ensuring farmers are not exploited by the sugarcane companies. However, their role in extension in terms of technology transfer is limited due to the influential presence of the company in the region.

DESCRIPTION OF THE EXTENSION APPROACH

The information value chain for sugarcane production in the study region is shown in Figure 4.2. Unless the company develops a lasting relationship with its regional sugarcane producers, the company's survival may be at risk for several reasons. Offering extension services is a natural conduit to connect with its supplier farmers and develop these partnerships. First and foremost, the company requires a regular supply of sugarcane for its crushing operations. The profitability of the sugar mills depends directly on the number of days per year that it is in operation. It cannot afford to keep the sugar

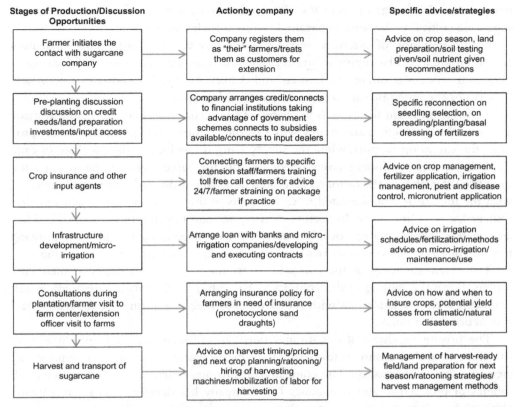

Stages of Production/Discussion Opportunities	Action by company	Specific advice/strategies
Farmer initiates the contact with sugarcane company	Company registers them as "their" farmers/treats them as customers for extension	Advice on crop season, land preparation/soil testing given/soil nutrient given recommendations
Pre-planting discussion discussion on credit needs/land preparation investments/input access	Company arranges credit/connects to financial institutions taking advantage of government schemes connects to subsidies available/connects to input dealers	Specific reconnection on seedling selection, on spreading/planting/basal dressing of fertilizers
Crop insurance and other input agents	Connecting farmers to specific extension staff/farmers training toll free call centers for advice 24/7/farmer straining on package if practice	Advice on crop management, fertilizer application, irrigation management, pest and disease control, micronutrient application
Infrastructure development/micro-irrigation	Arrange loan with banks and micro-irrigation companies/developing and executing contracts	Advice on irrigation schedules/fertilization/methods advice on micro-irrigation/ maintenance/use
Consultations during plantation/farmer visit to farm center/extension officer visit to farms	Arranging insurance policy for farmers in need of insurance (pronetocyclone sand draughts)	Advice on how and when to insure crops, potential yield losses from climatic/natural disasters
Harvest and transport of sugarcane	Advice on harvest timing/pricing and next crop planning/ratooning/ hiring of harvesting machines/mobilization of labor for harvesting	Management of harvest-ready field/land preparation for next season/ratooning strategies/ harvest management methods

Figure 4.2 Information value chain for sugarcane production. *Source: Author's compilation based on field interviews.*

mill idle while other competitors are increasing the supply of sugar in the market. Second, the area under sugarcane cultivation is decreasing because of urbanization and because subsidy programs for planting of cashews are diverting traditional sugarcane land to cashew farms. An increasing demand for bricks for construction is also diverting land away from sugarcane cultivation. Third, the yield of sugarcane has decreased from 40 tons per acre in the beginning of the 1980s to 25 tons per acre in the 2000s. The decline in productivity in the area means a low supply of sugar to the sugar mill. Part of the reason for the decline in sugarcane yield is a lack of groundwater and the high cost of irrigating the sugarcane fields. Although electricity is practically free, it is not supplied regularly, and farmers resort to diesel pumps to irrigate their fields, which is relatively expensive. Finally, the relative price of daily labor (for planting, intercultivation, and harvesting) greatly increased with the introduction of the NREGA employment guarantee schemes, which assure every poor household 100 days of employment per year at a daily wage of up to Rs. 100. This shifted labor away from agriculture and sugarcane cultivation because sugarcane farming is such arduous work.

Private provision of extension services is one way to address the above challenges and to ensure a regular supply of sugarcane for the crushing operation. A full range of extension services is offered to farmers and delivered under a needs-based approach. Services not required by the farmers are not forced upon them.

The vice president for canes is in charge of sugarcane procurement. A general manager runs the production and supply wings of the company. The production wing provides the extension support, while the supply wing is involved in the purchase of cane from the farmers, including provision of support for harvesting and transporting the cane from the farm to the factory. The production wing is divided into several divisions, with the divisional offices located in several parts of the command area. Each Division Officer has six Cane Sugar Inspectors (CSIs), and each CSI has three Sub CSIs. They are given a certain area to cover, about 1,200 acres depending on the cane-growing blocks. The sugar mill covers an area of about 80,000 acres for procuring sugarcane.

The extension wing of the company treats the contracted farmers as their customers. From the signing of the contract to harvesting of the crop, the extension staff maintains a close relationship with the farmers to meet the farmers' information and production input needs.

The private provision of extension services to enhance household production of sugarcane (and thus ensure a steady supply of high-quality input for the company) has created a win–win situation. The company aims to make farming constraint-free in all spheres of production and marketing. The company has developed a system of cane cultivation and management that makes sugarcane growing simple and cost-effective by connecting farmers to all the services they may need in production. Farmers get regular advice on production, from land preparation to harvest and delivery. The farmers are assisted with soil testing, equipment rental, and labor as needed. They are supplied with fertilizers and chemicals as necessary by the company's own fertilizer and chemical sister companies or through their authorized dealers. The company also helps with spraying biological control agents if requested by a farmer. Yet farmers are not forced to use these services. There is a call center open 24 h a day to receive the farmers' requests, and the company has devised a system to systematically follow up on such requests. Management later follows up on the requests to ensure the farmers were satisfied with the support provided by the company's extension agents.

In addition to supplying agricultural inputs, the company also facilitates access to credit through its finance department, including financing for investments such as digging bore wells, installing drip irrigation systems, and other land development investments. These loans are repaid by the farmers under individual arrangements between the farmers and the company. For larger loans, the company may stand as a guarantor for the farmer so that the farmer can obtain a loan from a formal financial institution.

Sugarcane cultivation is a labor-intensive operation. Labor shortages are becoming chronic in the study region, and the company has taken proactive steps to hire "labor gangs" from areas with labor surpluses, transporting workers to the farms as needed. The labor gangs usually consist of 20–30 laborers who move around the region, helping farmers to harvest their sugarcane.

The company has signed agreements with equipment companies, such as New Holland and John Deere, to supply needed equipment to its farmers. The company also identifies progressive farmers who would like to become entrepreneurs; it then helps them to buy equipment that can be rented to other farmers in the area. The company assists the entrepreneurs in taking out long-term loans to buy this equipment by acting as a guarantor. Additionally, the company identifies farmers who need equipment rental services and connects them to the entrepreneurs. These arrangements help farmers to avoid local money lenders, who often charge exorbitant rates of interest.

The company also runs a farmer training and knowledge center where modern methods of sugarcane cultivation are demonstrated to farmers. The center produces sugarcane seedlings for purchase by both contracted and non contracted farmers.

During the field work for this report, several farmers were interviewed in depth to better understand their interactions with the private extension services. The first farmer interviewed was a woman farmer who manages an 8-acre farm, on four of which she regularly cultivates sugarcane. She uses the traditional irrigation system, pumping water from an open well. She practices ratooning for 2–3 years in a row. She signed up with the company at the beginning of the season to sell sugarcane to the company and has visited the extension center to purchase seedings and to routinely seek advice from the company. She employs her own laborers to manage the sugarcane fields. She felt that the extension services provided by the company enabled her to take up cultivation of sugarcane without much help from her husband, who has a regular job. She felt the timely advice given by the company and the help provided to obtain the inputs and the laborers to harvest the sugarcane were the most useful services of the private company.

The second farmer was a small-scale farmer with only 2.5 acres of sugarcane. He has a good supply of water and has been growing sugarcane for more than 25 years. He has continued to grow sugarcane because his yield levels have been consistently high. He follows the advice of the company and their recommended schedule as closely as possible. He has rented a tractor through the services facilitated by the company through its entrepreneur development program. He attends the classes offered by the company at its training center two or three times during the crop season. As part of his contract, he receives fertilizer and chemicals from the company. According to this farmer, a major advantage of the private company is the timely access to inputs. He had found it difficult to obtain the needed inputs when they were needed during the crop season. Due to the small area cultivated, this farmer by himself was not able to buy a power tiller or a tractor. However, he was able to hire the local equipment renter organized by the company to plow the field and prepare the land for cultivation. This was very helpful for the farmer since labor availability has become a serious challenge to sugarcane production in the region.

The third farmer was a large-scale farmer who grows sugarcane on about two-thirds of his 18-acre farm. He also grows rice, groundnuts, and cashew nuts because the government provided a subsidy following a storm that wiped out the cashew trees. He feels that without the help of the company in cultivating sugarcane he would have stopped growing it. He has a drip irrigation system installed in his garden, which he obtained partly through a company-financed loan and partly through a subsidy from the government. He practices ratooning, which requires land preparation only once every 2 years. This large farmer was able to buy a tractor through credit arranged by the company with the local commercial bank. He was able to lay down a micro-irrigation system for 10 acres of his land with the help of the public extension system. He gets a 50% subsidy from the central government, 25% given by the state government. This farmer's main benefit from the private extension program is the use of labor-saving equipment and timely advice from the extension agents.

ANALYSIS OF THE PRIVATE EXTENSION SERVICES WITH RESPECT TO RELEVANCE, EFFICIENCY, EFFECTIVENESS, EQUITY, SUSTAINABILITY, AND IMPACT

Table 4.1 presents various interventions by the sugarcane company at different stages of crop production. Sugarcane farmers in India are usually under some form of contract with the processing company situated in close proximity to the farms. Sugar factories are run by the private sector or by cooperative societies of which the supplying farmers are members. Historically, sugar factories had to buy cane from within a designated area, but under recent amendments these laws were relaxed to allow farmers to choose the factory they want to sell to. The challenges of labor shortages and water scarcity, as well as a general trend toward decreasing dependency on agriculture, have resulted in crop diversification, especially in coconuts and cashews. These concerns have led the sugarcane factories to rethink the way they source their inputs, especially in terms of the services they provide to farmers. In the past, companies relied on public research and extension systems to support and educate farmers. However, over the past 20 years, the decline in both the quality and quantity of public extension services necessitated intervention by the private companies.

The private company is keen to ensure the supply of cane for at least 300 days each year. This requires that their extension services not only meet the farmers' information needs, but also their input and credit needs. This forced the private company to invest in its extension functionaries and hire additional staff to provide the farmers

Table 4.1 Inclusive innovation during various stage of crop production

Input provision	Extension support	Timing	Production advice	Harvesting/transportation	Price payment/financial arrangements
Credit institution linkages	Advice on where and when to get credit/repayment contracts	Pre-planting/planting	• Nursery development • Green leaf manuring • Soil testing • Land preparation • Basal dressing • Weeding/intercultivation	Advice on timing of the sugarcane harvest	Credit paid with interest to credit institutions
Public extension and research	Connecting farmers with public system of research and extension	30–60 days	• First top dressing • In-situ plowing • Weeding	Renting of the harvest machine from government depots	Farmers are guided on government programs supporting sugarcane farmers
National subsidy programs	Installation and use of subsidy programs for micro-irrigation from machinery	60–90 days	• Second top dressing	National agricultural development program activities and knowledge sharing	Subsidy costs are recovered and paid to government from harvest
Seedling supply	Disease-free seedlings for planting and supply of seedlings	90–150 days	• Third top dressing • Trash removal • Parasite card placement (biocontrol) • Pheromone placement (biocontrol)	Preparing land for next season; promotion of rationing	The price of the seedlings paid to nursery farmers/reach station/company
Fertilizer/pesticide/weedicide/supply	Advice and connecting to fertilizer and pesticide dealers	150–180 days	• Parasite card • Tetraticus control	Feedback on pesticide/chemical use	Cost of inputs are deducted and the input dealers paid
Mechanization for land preparation	Advise on use of equipment	180–210 days	• Trash removal • Pheromone placement • Tetraticus control	Advice on renting machinery for harvesting	Farmer/entrepreneur settlement/new contact for next season cost of harvesting paid to harvester leasers and labor gangs

Source: Author's compilation.

with comprehensive services. Given the competition for use of the land and the labor and water constraints, the extension services became not only highly relevant, but also highly individualized to farmers facing different constraints. The company makes continuous efforts to understand the information gaps of the farmers and to be alert to their needs in different aspects of agriculture. Thus, the extension services have become very specific to sugarcane production, but because of the holistic approach of the company, there are spillover benefits to farmers producing other crops (e.g., because of improved irrigation systems). Due to this effort, the public extension system has become less relevant to the sugarcane farmers to the extent that some of the families of extension officials have signed up with the company to supply sugarcane on a contract basis and receive the private extension services.

Discussions with the case study farmers indicated that the efficiency of the production system has improved in many ways. This is due mainly to the interventions and innovations that the company introduced to meet the day-to-day needs of the farmers, for example, in terms of information and inputs.

The change in the provider of extension services increased the effectiveness of the extension. This is because the farmers who are under a contract with the sugarcane company are followed up with regularly to ensure that their individual needs are met and that they are not motivated to divert their farmland to another crop. The need for focused attention on the farmers made the extension services provided to the farmer more individualized, and hence more effective.

Although the extension services are provided entirely by the company, it is difficult to tell to what extent the costs are borne by the company versus by the farmer. Because the company must provide services to retain the farmers' interest, since they have alternative agricultural options, it is suggested that more of the costs are absorbed by the company than in the typical scenario where the processing company has more power than the farmer. The provision of extension services (information and training) in combination with farm inputs also increases the cost-efficiency of providing extension. Finally, the company has found alternate ways to save money (e.g., by outsourcing equipment rental to farmers it worked with previously) or to make money (e.g., by selling seedlings from the demonstration farm at its training center).

This combined effort has increased the average yield of the sugarcane farmers from about 60 tons per acre to close to 80 tons per acre, resulting in an approximate 33% increase in the income of the sugarcane farmers. This income increase could be sustained if the extension system continues to meet the information and input needs of the sugarcane farmers.

The private company helped to bring many smallholder farmers who were moving away from sugarcane cultivation back, through the increased drive to sign contracts with them. One of the attractions of joining the contractual arrangement is the provision of timely inputs and extension advice from the company. This brought equity to the sugarcane farmers in accessing information on sugarcane cultivation. In the past the sugarcane company focused on large-scale farmers, concentrating its efforts to obtain economies of scale. However, the move toward provision of extension to all farmers brought with it a high level of equity among the sugarcane farmers.

The impact of private extension can be seen from several angles. First, there was an increase in the number of smallholders participating in and benefiting from sugarcane cultivation. Second, the yield levels of sugarcane growers under contract and receiving advice from the private extension workers have improved. A set of informal interviews with the farmers who produce sugarcane in the same area but not covered by the private extension indicates that there is an increase of yield (about 10–20 tons) in the sugarcane due to private extension. While these figures vary widely from farmer to farmer, the private extension does contribute to farmers' yields in various ways. For example, a major impact is reducing the transaction cost of obtaining inputs for sugarcane cultivation through an authentic source—the sugarcane company itself, which ensures that proper advice is given for their use and reduces the role of input dealers who may not sell the right inputs. Finally, the labor shortage that agriculture faces in the region is overcome by the company, which organizes the labor groups to work with the farmers in shifts. Overall, the private extension has had a considerable impact on the farmers, whose operations have been made hassle free.

The sustainability of the extension system is ensured in several ways. Since the company implicitly charges for the extension through exclusion of extension costs in the price paid to the farmers, private extension services can be sustained as long as the farmers are in contractual arrangement with the sugarcane company. However, long-term sustainability depends on the survival of the sugarcane company itself. The demand for sugar and sugar products from the company and the profitability of the sugarcane company will in the long run determine the level of engagement of the company with the farmers. Further, given the need for expansion of sugarcane cultivation to maintain an adequate supply of sugarcane to the crushing operation, the company has a vested interest in providing the extension service as an additional incentive for the farmer to grow sugarcane. This creates a win–win situation for the farmer and the company. As the company keenly identifies the production, information, and credit bottlenecks of the farmers who are under contract with them, farmers are more likely to grow sugarcane. To some extent this has created a level of dependency among the farmers, who are beginning to question the implicit price they pay for these services. The system of private extension can be sustained as long as a safe balance is maintained in terms of the dependency and the price they indirectly pay to the company for the extension and other services the company provides.

Perspectives of the producers/farmers

Farmers have been facing a number of challenges in increasing the yield of sugarcane over the last 20 years. This company was the only company to which they could sell the sugarcane. The company buys the sugarcane only when there is a need for sugarcane. The timing of the sugarcane harvest was such that all the farmers in the area had sugarcane ready for harvest at the same time. The company did not have an incentive to increase the yield of the farmers since there was plenty of sugarcane to be harvested at the same time. So they would stagger the cutting depending on the operation of the crushers. However, the company's new objective to keep the mills running throughout the year meant that sugarcane needed to be harvested all the time, and with micro-irrigation systems farmers can plant sugarcane throughout the year.

Perspectives of the public extension workers

Individual and group interviews were conducted with public extension workers and officials in the study area to assess the status of public extension after the introduction of intensified private extension. The feedback from the public extension officials showed mixed opinions on the role of private extension. Most of them agreed that the private company is in a better position to provide—and does provide—a range of services including inputs, credit, micro-irrigation, and advice on everything from land preparation to post-harvest, offered according to the needs of the farmers. This was an attractive feature of the private extension that the public extension system could not provide. Some of the public extension officials were also contacted by the sugarcane farmers and have reported that they received high-quality services from the company. This has exerted a certain level of pressure on the public extension system, whose mandate is to serve the farmers in the area. But due to limited staff and resources, public extension officials are fine with the role taken by the sugarcane company to provide the extension services. In addition, the public extension system is still responsible for providing support to the farmers in the region who grow other crops, such as groundnuts, rice, and other perennial crops such as cashews. Further, the public extension continues to implement the subsidy-based extension operation. This helps to complement the private extension to some extent. However, a major discontent among the public extension officials is that the company does not price the sugarcane taking into account all the by-products produced from sugarcane crushing, but rather bases the price paid to farmers only on the sugar content and its quality. Some officials saw the connections between the private company and the input dealers and credit providers as a system that collectively exerted too much control over the farmers. In addition, the company's recommendations for fertilizer and chemical use varied from that of university researchers, which the public extension staff followed. There was a feeling among the officials of the public extension system that since the chemicals and fertilizers were also supplied by the company at a predetermined price, there could be over-application of fertilizer. Soil testing rates for the samples taken by the company were higher than the soil testing done at the government laboratories. Some public extension workers saw the privately provided extension services as a threat to their work, and they felt that the farmers whom they served previously did not take them seriously anymore.

Perspectives of the staff of the company

Staff members of the company were generally confident about the benefits of the services they provide to the farmers. They saw themselves as a solution to the farmer's day-to-day problems, and they were willing to go the extra mile to meet farmers' needs. They felt that they were accountable to the company if the farmers had low yields. They conducted routine surveys of farmers' satisfaction to find out how they could better help the farmers.

COMPARISON WITH PUBLIC EXTENSION IN INDIA

How does the private company offering extension services differ in its approach compared to other service providers in the region? Besides the private company, a major source of extension services for the sugarcane growers in the region is the public extension system, which by default is supposed to be there for the farmers irrespective of the type of contract the farmers have with any company that procures sugarcane from them. Yet the services provided by the public system are not highly efficient, partly due to the pullback of the public system, which was overshadowed by the private extension and partly due to the general decline in responsiveness of the public extension to the demand for information and services by the sugarcane farmers. In addition, there are other input dealers who provide partial service to the farmers, mainly in terms of chemical inputs and pest control.

Some of the features that make EID Parry more effective in delivering extension services include the following: First, the company has direct contact with the farmers. As soon as a farmer contacts the company to become a seller of his or her sugarcane to the company, the company registers the farmer and makes the farmer part of its database to follow up on all aspects of farming. Second, the farmer can get the services that the company offers but still ends up paying the market price for the inputs. This means that the farmers do not incur any additional cost for the inputs they receive from company other than the price that prevails in the market for the inputs. Third, this service orientation distinguishes the company from others who may provide extension service in the area. The company's objective is to keep the farmers hassle free and to make them part of their production system. To achieve this objective the company goes the extra mile to identify the general and specific challenges facing the farmers and develops solutions. Fourth, the open line of communication through the dedicated phone line which responds to farmers need within 48 h is a key to increasing the dependability of the company and in encouraging the farmers to trust in the company as the source of advice. Finally, the extension center that is open every day of the week allows farmers to stop by and learn of recent technological developments in sugarcane production.

UNIQUENESS OF THE MODEL AND VALUE ADDITION (INSTITUTIONAL INNOVATION)

The institutional innovation for the knowledge work of the company comes from the combination of several roles into one package deal. This reduces the challenges that the typical smallholder farmers faces multifold. Before deciding to plant a particular crop, the uncertainty of the price of the outputs is removed by the range of prices that the farmers receive for their sugarcane.

The policies and programs of the government during the last decade have focused on breaking the subsistence nature of farming and helping smallholder farmers to become agricultural entrepreneurs who operate their farms as businesses. Yet the challenges of improving the business skills of smallholder farmers persist, including the fragmentation of produce markets, lack of technical knowledge around commercializing farming practices, poor access to quality inputs, and inability to obtain credit at competitive interest rates. Further, unreliable markets for outputs and middlemen at various stages of the supply chain make it difficult to operate efficiently. This is the so-called "smallholder syndrome."

The smallholder syndrome is broken by the company through its provision of extension and advisory services. This is the unique aspect of the model. How does the company decide where to get the expertise it needs? The company effectively pulls together the global, regional, and national knowledge needed for successful cultivation of sugarcane by the area farmers.

What knowledge model does it use? The company believes that if the farmers are adequately equipped with knowledge to grow sugarcane and if this is backed up by provision of services in getting adequate quantities of inputs at the needed time, then farmers will stick to growing sugarcane. Making sugarcane cultivation a trouble-free enterprise is the operational goal of the extension wing of the company.

LESSONS FOR REPLICATION AND SCALING UP

A major challenge in increasing productivity, even in areas where inputs are readily available, is the lack of awareness with regard to recommended practices in growing particular crops. Private provision of extension could be a sustainable alternative to a failing public system if it gives the private provider a competitive advantage over other companies in producing the final output. Public policies could enhance the role of the private sector, at least in areas where private companies benefit from working with farmers for input supply and output aggregation. The command area policy, which restricted cane procurement to a specified geographical area, helped to establish this linkage over four decades ago. Yet the relaxation of this policy now can increase competition among the sugar mills for the produce, and the farmers will be the ultimate beneficiaries of this competition.

The integration of input supply and extension services can complement each other in a mutually beneficial fashion. The sugarcane production system studied here shows that inclusive innovation is possible under specific circumstances that bring various players and actors together, not by chance but by careful programming and policy-making. Policy support to the farmers and innovation entities may have a significant effect by stimulating inclusive innovation approaches. However, such approaches need to be supported by strong investment in knowledge management and sharing. Adaptive innovation and increased competition between service providers can result in increased inclusiveness of the innovation system. Private extension is not a panacea for all farming systems and all the smallholder farmers who still are not included in the innovation systems. However, policies that guide regulatory interventions could help identify opportunities to enhance the role of private extension providers. Implementation gaps exist at all levels, and their removal could enhance the role of private extension. A key lesson from this case study thus is the need for better understanding of the role of public policy in shaping the inclusive innovations that could be supported by private extension for adoption by poor farmers (Foster and Heeks, 2013a,b).

CONCLUSIONS

In this chapter, we undertook a case study of the extension services provided by a sugarcane processing company in India to understand how a private-sector output aggregator provides extension to its farmers. This situation is unique, because only 10 years ago the company dictated prices and quality standards to the farmers, and has now had to turn around to provide support to farmers to keep producing sugarcane for its operation. Both the water scarcity and labor scarcity due to local development of nonfarm activities have forced farmers to move from sugarcane to less labor- and water-intensive perennial crops. To entice farmers to stay in sugarcane farming, the company has focused on removing these constraints and making the sugarcane production hassle free for the farmers. This resulted in a win–win situation for both the farmers and the sugarcane crushing company. However, unless output aggregation continues to be accompanied by an effective extension and farmer support system, the sugarcane industry may yet face an unsustainable future.

REFERENCES

Anderson, J.R., Feder, G., 2004. Agricultural extension: good intentions and hard realities. World Bank Res. Obs. 19 (1), 41–60.

Anderson, J.R., Feder, G., Ganguly, S., 2006. The Rise and Fall of Training and Visit Extension: An Asian Mini-Drama with an African Epilogue. World Bank Publications, Washington, DC.

Babu, S.C., Joshi, P.K., Glendenning, C., Asenso-Okyere, K., Rasheed Sulaiman, V., 2013. The state of agricultural extension reforms in India: strategic priorities and policy options. Agric. Econ. Res. Rev. 26 (2).

Birner, R., Anderson, J.R., 2007. How to Make Agricultural Extension Demand-Driven? The Case of India's Agricultural Policy. International Food Policy Research Institute, Washington, DC. Discussion Paper 00729.

Foster, C., Heeks, R., 2013a. Conceptualizing inclusive innovation: modifying systems of innovation frameworks to understand diffusion of new technology to low-income consumers. Eur. J. Dev. Res. advance online publication, 4.

Foster, C., Heeks, R., 2013b. Analyzing policy for inclusive innovation: the mobile sector and base-of-the-pyramid markets in Kenya, innovation and development. Eur. J. Dev. Res. 3 (1), 103–119.

CHAPTER 5

Kenya Horticultural Exporters: Linking Smallholders to Market

Yuan Zhou

Syngenta Foundation for Sustainable Agriculture, Basel, Switzerland

Contents

INTRODUCTION

The Kenyan economy is heavily reliant on the agricultural sector, which accounts for nearly one-third of national GDP. Horticulture is an important subsector of Kenyan agriculture. It has grown significantly over the last few decades to become a major source of employment and foreign exchange earnings. The subsector comprises a mix

Knowledge Driven Development.
DOI: http://dx.doi.org/10.1016/B978-0-12-802231-3.00005-X

of products: flowers, fresh fruits, and vegetables. Farms involved in this activity vary widely in terms of farm size and geographical area of production. The horticultural industry has grown from being primarily small businesses and smallholder farmers to being dominated by sophisticated businesses that are increasingly vertically integrated (SNV, 2012). This transformation presents new opportunities and challenges in addressing the needs of farmers in this value chain.

Kenya has been exporting vegetables to Europe since the 1950s. The industry has faced and successfully overcome many challenges such as freight bottlenecks, new competition, macroeconomic instability, and new requirements from regulators and customers. Kenya's original success in exporting vegetables, especially beans, was due to its climatic and geographic comparative advantages. Year-round vegetable production and the availability of northbound airfreight made exporting horticultural products lucrative for Kenya. Its sustained success has been due to market segmentation, investing in certification schemes, adding value to products through sophisticated packaging, servicing niche markets, and investing in marketing (SNV, 2012). Despite a history of inefficient and inconsistent provision of extension services, the Kenyan government has supported critical areas of the industry through investment in education, provision of phytosanitary services, facilitation of expanded freight services, and increased access to the latest technology.

Regulation of the international fresh vegetable market poses new challenges to growers, especially smallholders. Growers have to meet standards for environmental management, food product safety, quality, traceability, and workers' health and safety. In 1996, the Fresh Produce Exporters Association of Kenya (FPEAK) developed a code of practice (Kenya-GAP) to help producers and exporters meet international standards. Over the years, the adoption of many international standards, such as GlobalGAP, Fairtrade, BRC, and individual customers' standards, by both large and small producers have facilitated an expansion of the export industry. Most recently, issues around maximum residue limits (MRL) and the subsequent trade restrictions imposed by the EU have brought additional challenges to the sector.

Many exporters work with both large and small farms to ensure a reliable supply of high-quality products. Their outgrower schemes are often systems of contract production that source directly from a large number of small-scale farmers. Exporters provide critical inputs, specifications, training, and sometimes credit to their suppliers, who in turn provide assured quantities of specialty produce at guaranteed prices. Kenya Horticultural Exporters (KHE) is one of the leading firms that have an extension model for its outgrowers, a large majority of whom cultivate French beans.

The following analysis will focus on the French bean sector, describing the exporters' extension services to smallholders. The relevance, efficiency, effectiveness, equity, sustainability, and impact of this private extension program will be discussed, as well as lessons for scaling up and replication.

THE BEAN VALUE CHAIN

In Kenya, French beans are primarily grown in Central province and parts of Eastern and Rift Valley provinces. The vegetable was originally cultivated exclusively for export, but over the years it has become popular in national markets. The crop is grown under irrigation, mainly by smallholders. The main importing countries are the United Kingdom, France, Germany, the Netherlands, Belgium, and South Africa (HCDA, 2010).

Key trends in production and export value

Between 2008 and 2010, production volume and value of the beans fell considerably (Table 5.1). This was due to a prolonged drought in 2008–2009. In 2010, the farm gate value amounted to Ksh. 1.6 billion (1 Ksh. = 0.011 US$), whereas the export value stood at Ksh. 4.4 billion. This was because of the relatively low farm gate prices offered to farmers by exporters, who then made large profits from value-addition activities such as processing and transport. In 2012, French beans accounted for 38% of Kenya's total earnings from vegetable exports of Ksh. 20 billion (HCDA, 2013). Since 2006, the percentage has varied between 26% and 56% (Figure 5.1).

An estimated 50,000 smallholder farmers, who operate up to 2 acres (1 acre = 0.4 hectare) of land, are responsible for growing most of Kenya's French beans (SNV, 2012).

Table 5.1 French bean production and export statistics, 2006–2010

Year	Area (ha)	Production (metric tons)	Farm gate value (million Ksh.)	Export value (million Ksh.)
2006	6,154	61,540	1,846	8,726
2007	7,733	67,330	2,020	7,858
2008	4,616	92,095	2,925	5,579
2009	3,336	46,496	1,739	4,306
2010	4,840	55,841	1,606	4,400

Source: From HCDA (2010).

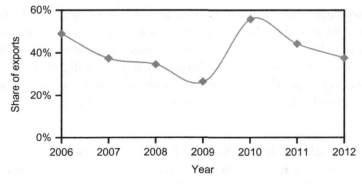

Figure 5.1 French beans' share of total vegetable exports by value. *Source: From HCDA (2010).*

They grow horticultural crops because they are profitable, earning up to seven times more income than maize on average. Family members provide most of the labor. Some of the smallholders additionally work on large farms. The French bean industry employs 45,000–60,000 people in commercial farms, processing, and logistics operations.

The value chain partners of the bean sector

The bean value chain in Kenya is increasingly complex, with a multitude of players. In the export chain, there are farmers who grow the crop, middlemen and exporters who contract or procure from them, airlines and logistics companies that ship the product, the government as a regulator, and supermarkets in Europe where the crop is sold.

Growers

There are several distinct types of growers, namely large commercial farmers, small to medium farmers who grow beans primarily on contract, and very small traditional or emerging growers. Large farms typically grow vegetables as a business, with tight cost controls and scientific planning. They are usually owned by or associated with exporters or processors. Contract farmers are usually organized into groups, under which they cultivate as individuals and jointly market the collected beans. Through an agronomist the exporters provide inputs and technical assistance to a group, typically of about 30 farmers. To make operations economical, farmers need to be close together and each has at least half an acre under bean production in order to supply at least 1–2 tons per collection (SNV, 2012). The third group is farmers who work with exporters but may not have a written contract. These include very small farmers who are not organized. Producing high-quality beans and following standards poses a challenge to them, and the loose relationship often causes side-selling.

Collection and brokers

Because of increasing demands for traceability of produce, exporters have to directly contract growers. This has reduced the overall need for brokers, but exporters still frequently use middlemen to fill shortfalls in their own production pool. In farmed areas not yet reached by exporters, brokers are the growers' main outlet, but they pay relatively low prices. Brokers can source from a range of growers and sometimes buy directly from other companies' contract farmers by offering a slightly higher price. This is called side-selling or poaching, and it is a major problem for exporters as they have invested in inputs and extension services without reaping the benefits.

Exporters

Exporters' roles include purchasing the beans from growers, grading and packing, as well as shipping to a buyer in Europe. There are three major groups of exporters: the large, small to medium, and "briefcase" exporters. The large exporters have increasingly integrated their operations, both forward into the markets and backward into production.

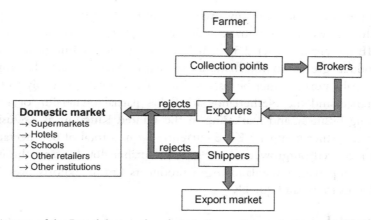

Figure 5.2 Diagram of the French bean value chain in Kenya. *Source: Author's compilation.*

They usually have their own large farms as well as outgrower schemes, where they contract and train small farmers to produce high-quality and safe vegetables. Sunripe, KHE, Finlays, Vegpro, and Everest all belong to this group. Small and medium exporters are serious about exporting, but do not have the resources or ability to reach a large scale. They typically run smaller farms of their own and work with contract growers—large groups of farmers or large individual farmers. Briefcase exporters operate only during the high season when prices are high. They procure from brokers, loose-pack the produce, and ship to buyers. As they do not have regular growers and therefore face difficulties with traceability, they tend to play a smaller role in the market.

As French bean production is not primarily targeted at the local market, the distribution channel differs from those of other common vegetables (i.e., from farmers to wholesalers to retail outlets). Products unsuitable for export are ultimately consumed via domestic supermarkets, hotels, schools, and other retailers or local institutions. Figure 5.2 shows a diagram of the bean export value chain.

A number of key institutions, both public and private, support the growth of the horticultural sector. The major public actors include the Ministry of Agriculture, Horticultural Crops Development Authority (HCDA), Kenya Plant Health Inspectorate Service (KEPHIS), Kenya Agricultural Research Institute (KARI), and the Pest Control Products Board (PCPB). Private-sector actors include FPEAK, the Agrochemical Association of Kenya (AAK), the Seed Traders Association (STAK), and the Kenya National Federation of Agricultural Producers (KENFAP).

EXTENSION METHODS OF KHE

KHE Ltd. is one of the country's leading exporters of high-quality vegetables and fruits. Established in 1977, KHE has steadily expanded its range of produce and geographical presence. The company owns two large farms in central Kenya, Ontilili and Bahati. It also works with many independent producers through its outgrower schemes. KHE has enabled over 2,500 small farmers to access European markets with first-class produce grown to international standards. Major vegetable crops include French beans, baby corn, runner beans, and broccoli. Exports go mainly to the United Kingdom, France, and the Netherlands. KHE has partial ownership of a major air-freight handling business, and marketing in the United Kingdom is exclusively done by Wealmoor, a partner company. KHE currently farms a total of 690 hectares directly and through managed outgrower schemes, and a further 460 hectares through a contractual partnership with Greenlands Agro-producers Ltd. It exports over 5,500 tons annually and aims to expand quickly.

Extension model

KHE plans its vegetable production and operations in line with clients' requirements. In its outgrower schemes, the company assesses individual supply capacity by sitting down with farmer groups or cooperatives to understand their land and labor availability. KHE's products follow three major safety standards: GlobalGAP, Leaf, and Tesco-Nature. There are generally two sources of vegetable supply: KHE's own farms or contract farmers. The share of products coming from contract farming varies, but averages around 60% in a normal year. However, since January 2013 the market has been especially challenging due to stringent EU requirements on traceability and MRLs. As a result, KHE now plans to source only 40% of total production from contract farmers and supply the rest through its own farms.

Outgrower scheme and extension services

In its outgrower system, KHE currently works with 200–300 cooperatives and 20–30 farmer self-help groups (SHGs) or business groups around Mount Kenya. This region has been chosen because of its favorable climate, proximity to Nairobi, and relatively experienced farmers. KHE usually signs contracts for a year, renewable if a trusting relationship develops. The contracts clearly lay out the roles of KHE and growers, as well as acreage, prices, dispute resolution, and aspects of ethical trading. The basic role of KHE includes:

a. Provision of a supply program
b. Provision of instructions on food safety management
c. Monitoring program implementation through staff visits
d. Provision of product specification in line with market requirements
e. Purchase of produce at agreed prices
f. Collection of produce wherever possible

Farmers are expected to follow the agreed planting program and the company's technical recommendations, implement good agricultural practices (GAPs), and

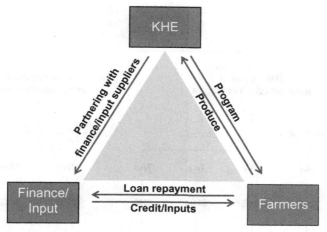

Figure 5.3 Extension approach of KHE. *Source: Author's compilation.*

comply with all food safety measures. KHE provides a list of approved crop protection products (CPPs), which farmers must strictly follow. As part of the process, the company also assesses groups' capacity in terms of acreage availability and teaches them how to run the program. Individual farmers fill out an application form to become a contracted group member.

The key services provided by KHE are provision of seeds (deductible from sales proceeds or paid in cash), arranging provision of fertilizer and CPPs, developing the planting schedule/program and guidelines to meet, and training on GAPs and compliance with food safety and quality standards. A simplified version of KHE's extension model is illustrated in Figure 5.3. In the ideal case, KHE provides seeds on credit and arranges for financial partners to provide loans to its contract farmers for purchasing other inputs such as fertilizer and chemicals from selected providers. However, involving a third party such as a financial institution can be difficult because of the risk that farmers might sell outside of the contract. As a result, KHE uses this approach only with the groups regarded as loyal, which account for about 40% of the total number of contract farmers. The rest have to pay for the inputs upfront in cash. There are two particularly critical areas to get right here: first, to find credible sources of CPP for the farmers and, second, to have a good quality management system (QMS).

Outgrower management model

KHE's outgrower management model includes a team of agronomists to provide farmers with technical assistance and quality assurance, with a clear division of responsibilities at multiple levels (Figure 5.4). For example, KHE's Group Agronomist is responsible for developing and reviewing the QMS and managing the overall system as well as establishing relationships with local and international development agencies.

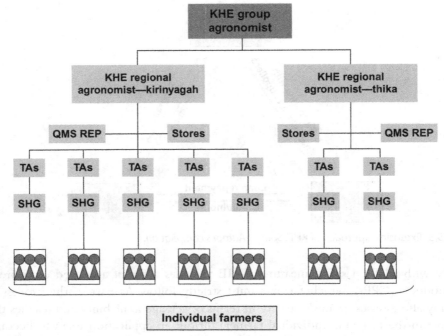

Figure 5.4 Outgrower management model by KHE (*note:* QMS REP = QMS representative). *Source: Author's compilation.*

The KHE regional agronomists are accountable for implementation and overall supervision of production in their regions. QMS representatives supervise a limited number of groups, ensuring systems are adhered to, and provide technical support to the technical assistants (TAs) and farmers. They are responsible for all CPP and fertilizer applications as well as ensuring that only KHE-approved products are used. TAs supervise planting, CPP, fertilizer application, scouting, and record-keeping for individual farmers. The farmer SHGs provide funds for infrastructure, collection centers, and input procurement. Farmers are responsible for cultivation and harvesting processes, with guidance from TAs in the SHGs.

Cost recovery

It is hard to quantify the exact cost of extension services. The overall cost of private extension includes overhead and operations cost. The overhead cost consists of personnel and administration; operations include such items as transport, logistics, training, and communications. Operations cost is needs-driven and highly variable from year to year, especially in an unpredictable climate and market environment. Normally when KHE is offered a price for a certain crop from importers, it calculates back the raw material cost (farm gate crop price), packing and processing costs, and operations costs, aiming for a profit margin of 15–25%. However, this margin has been under pressure in recent years because farm production costs have increased while European price offers have remained relatively steady. The extension costs are embedded in the farm gate prices.

Farm-level engagement and value created by KHE

We had the opportunity to visit three farmer organizations that work with KHE, including one cooperative and two farmer groups.

Murinduko irrigation cooperative society limited (MICSL)

Murinduko is a farmer cooperative with 487 members, one of six KHE partner in the Embu region. KHE has been working with the group since early 2012, when the cooperative acquired irrigation through a government-supported water project. KHE has a seasonal contract with the cooperative, and interested farmers can enroll in the program to grow French beans.

Farmers typically hold about 3 acres of land and use ½ to 1 acre for KHE vegetables, mainly French beans and baby corn, all under irrigation. The productivity of French beans is about 7–9 tons per acre, while baby corn yields about 4 tons per acre. The rejection rate by KHE is around 10%. Farmers sometimes hire labor for harvesting, which costs about Ksh. 250 per person per day. KHE is currently offering about Ksh. 55 per kilogram due to increases in farmer production cost, about Ksh. 5 more than last season. Farmers earn about Ksh. 120,000 per acre in net income per season. However, they face a major disease problem in rust, for which it is often hard to get the right sprays.

Two KHE agronomists and two bookkeeping clerks work with this cooperative. The ideal ratio of agronomists to farmers for KHE is about 1:200. In addition, about 20 spray men are needed for a group of this size. Since the cooperative is relatively new, it has not yet been GlobalGAP certified. For the current season KHE provides MICSL with vegetable seeds and chemicals on credit, deductible from sales proceeds at the end of the season. In addition to the credits, KHE also fully records each farmer's activities such as planting, fertilizer application, sprays, and harvesting, using the company's QMS.

Gatugura farmers group

This SHG is located in Gatugura village, Kirinyaga County. It has worked with KHE for the last 6 years. There are about 60 farmers in this group. Farmers typically have up to 1 acre of land and on average devote half an acre to French bean production. The acceptance rate of produce from this group is quite high, nearly 100%. Farmers harvest three times per week. A good farmer with half an acre can harvest 300–500 kg per week and make Ksh. 200,000 per year. One farmer has even earned enough from beans to send his two children to university and college. KHE provides seeds and technical guidance. It collects the produce twice a week and pays the farmers every 2 weeks.

Green valley SHG

The Green Valley SHG is located in Ngusishi, Nanyuki. It has 18 members and has long been working with various exporters; it joined KHE in 2012. The group is currently in the process of forming a cooperative in order to be able to market its local vegetables independently.

ANALYSIS OF THE PRIVATE EXTENSION PROGRAM WITH RESPECT TO RELEVANCE, EFFICIENCY, EFFECTIVENESS, EQUITY, SUSTAINABILITY, AND IMPACT

Contracting small growers to purchase produce from them that is then input into a company's value chains is common practice in the Kenyan horticultural export industry. Most smallholders are poor, and their main role in the value chain is growing beans. Therefore, interventions that help these farmers improve the quality and quantity of production and get access to export markets are likely to have a significant effect on household incomes and food security. Given the suitable climate and small farm sizes, growing export-oriented crops is a good option for farmers to get the best value for their available resources. The introduction of export crops is very important for the regions involved. As shown above, growers generally yield good returns on their investment. If they do especially well, they are able to support the whole family and educate their children.

Since food safety standards are now a basic requirement for entry into export markets, smallholders need new skills and knowledge to be able to comply with the standards and provide details for accreditation. They are often assisted by exporters in doing so. Although these standards represent barriers to smallholders, they also bring benefits through improved mechanisms for managing small-scale suppliers, enhanced farm management skills, and generally increased GAPs (Owuor, 2008). Smallholders can bridge their technical and knowledge gaps through extension services provided by exporters. Through the SHGs, farmers have been able to learn and develop skills required to meet standards for food safety and hygiene, not just for export products, but also for the domestic market. Farmers have also increased their environmental awareness, especially in terms of waste management and pollution control. Additionally, exporters have established a traceability framework, developed quality management strategies, and organized the value chain to allow farmers to concentrate on farming and pest and disease management.

KHE has evolved the extension model to increase its efficacy and cost-effectiveness, and makes continuous efforts to improve the QMS. From the producers' viewpoint, the training and advice are specifically tailored to a certain crop, and can be easily understood and hence adopted. Every effort has been centered on achieving high productivity with safe vegetables.

The system is also equitable, because SHG farmers are entirely at liberty to decide whether to join the export activity. There has been no selection of farmers by KHE. Once on board, all farmers have access to the same services, regardless of gender, farm size, or other differences. In addition, as the prices and product specifications are determined prior to the season, the farmers are treated completely equally in terms of product inspection and payment processes. When recruiting farmers, however, there may be a bias against those with especially small farms—for example, less than a quarter of an acre—because it may not be economically viable for exporters to contract farmers for such a small amount of output.

With regard to sustainability, there are several aspects to consider. On the financial side, the extension model has proven successful and sustainable; the value added to export crops covers the costs of extension. As long as the export business thrives, the model can continue. Most export companies choose to work with both large farms and small growers to complement each other and hedge on supply. Working with small growers alone can be risky when the problem of poaching is prevalent or when the MRL requirement is formidably high. Another issue is trust between grower and exporter. Long-term relationships always build on mutual trust, loyalty, and win–win situations. Some exporters have had contractual relationships for 10 years with groups that do not engage in side-selling. Due to the increasingly stringent EU regulations for MRL, some exporters have downsized their outgrower schemes to reduce the risk of excess residue on the produce of individual farmers. The sustainability of such models depends on product demand and requirements of the export market, the domestic operating environment (e.g., the cost–benefit of production and regulation of side-selling), as well as competition from exporters from other African countries.

Participation in outgrower schemes generally has a very positive impact on farmers' incomes. Most farmers have chosen to grow French beans because they are more tolerant plants and easier to grow than runner beans or garden peas. Farmers can realize a return several times higher than with corn. In recent years, because of increasing production costs, farmers have wanted better prices for their French beans. However, exporters' profit margins have been squeezed because they have already increased the prices paid to farmers despite receiving unchanged offers from Europe. Currently, several companies are reportedly operating only at about break-even, with margins close to the bottom limit. At the same time, extension staff make great efforts to maintain farmers' profitability by helping them avoid or reduce product rejection.

COMPARISON WITH PUBLIC EXTENSION AND OTHER PRIVATE SECTORS IN THE COUNTRY

The public extension system in Kenya is now generally perceived as ineffective and outdated, and therefore unable to cope with the dynamic demands of modern agriculture. In the 1970s and early 1980s, agriculture contributed 38% of GDP; extension services were well resourced and focused on input supply and marketing. The ratio of extension staff to farmers was around 1:400. From 1984 to 1989, the country followed the World Bank's "Training and Visit" system. From 1989 onward, structural adjustments required the government to reduce extension staff and scale down its public functions. Kenya's agricultural extension budget and employee numbers plummeted, and the remaining staff had very little transport. Recruitment for extension only began again in 2006, with a resulting major gap between actual and needed expertise. A new extension strategy was adopted in 2003 with the aim of reaching more farmers. The public extension system currently employs about 5,400 staff, of whom 4,000 interact directly with farmers (personal conversation with MOA, Kenya). However, with a ratio of extension staff to farmers under 1:1,000, this number is far from ideal.

In June 2012 the Government of Kenya passed the National Agricultural Sector Extension Policy (NASEP) to improve service delivery. NASEP has a strong focus on pluralistic and demand-driven extension services. It also addresses funding modalities and regulations for extension services. Extension is increasingly seen as a complex system in which services are provided by a range of private- and public-sector entities. The services provided by private companies such as horticultural exporters are considered to be crucial in commodity-based extension.

Discussions with national extension and district agricultural officers revealed that a lack of manpower, financial resources, and transport still represent major bottlenecks. In the Laikipia District, there are an average of six agricultural officers, one of whom is dedicated to extension. And at the location level there is one officer only, who does not usually visit farmers but runs an information desk one day per week. Currently, the ratio of extension staff to farmers is around 1:850, far below the ideal figure of 1:200 recommended by the district officers. According to horticultural exporters, public services are not generally available where they work, and the technologies are often outdated. Exporters also point out that the Horticultural Crops Development Authority is supposed to regulate contracts, but fails to do so, making it even harder to prevent side-selling.

The export company Sunripe uses a smallholder management model similar to KHE's, with various levels of agronomists and technical assistants providing support to growers. Since current interest rates are quite high—about 18–20%—Sunripe is looking for cheaper options such as in-house credit. CPP suppliers Bayer and the SwissCoop supermarket support a pool to fund outgrower schemes. Like KHE, Sunripe is currently grappling with the EU's new MRL requirements and with poaching. In contrast to KHE, Sunripe has already expanded its production base to neighboring countries such as Tanzania and Ethiopia.

Among private companies, horticultural exporters are the most involved in extension activities. Strict quality, quantity, and timing requirements give marketers and processors of horticultural products an incentive to provide farmers with extension and supplies of input. All of the top companies in Kenya (Vegpro, Sunripe, KHE, Indufarm, etc.) have their own farms, and buy from farmers on contract or through middlemen. They provide extension services and inputs as well as standard compliance services to contract farmers. Small exporters are less able to afford to this level of engagement. Agrochemical, fertilizer, and seed companies are also involved in extension and technology transfer, albeit in a less holistic and integrated way. Most commercial distributors of inputs use sales representatives to carry out demonstrations or other technology transfer activities in the field, particularly when there is a new product to introduce. They usually collaborate with MOA extension staff who help organize meetings. Other extension activities include field days, radio programs, written brochures or posters, as well as training on safe use of chemicals.

UNIQUENESS OF THE MODEL AND ITS VALUE ADDITION

Contract farming is common in the Kenyan horticultural industry. Components of extension models and smallholder organization vary between crops. KHE's extension approach is representative of how most vegetable processors work with outgrowers. Its outgrower management model, combined with its QMS, has proved central to its success in contract farming. The model enables smallholders to capture value in export markets and empowers them to understand and work with international standards. Exporters also create further value for local communities by employing thousands of people in processing, packaging, and logistics.

There are many ways of conducting contract farming. Some focus only on sourcing from smallholders, while others are involved in the whole growing and harvesting process including provision of inputs. The most suitable model depends on local conditions as well as market requirements. The present case has proved successful in delivering service to growers and integrating them into the bean value chain.

LESSONS FOR REPLICATION AND SCALING UP

The extension model presented here is used with contract farming for a specific value chain crop. It has proved workable and effective in Kenya's vegetable export business. The same model could be replicated in other regions or countries for other high-value horticultural crops. As the success of the sector depends on serving a niche market and having good air connections to importing countries, the need for such an extension model in a specific value chain should be examined beforehand. Generally speaking, only high-value crops are likely to cover the cost of intensive extension without public-sector or third-party support.

The opportunity for scaling up such an approach in Kenya depends on several factors. First, if the export business keeps expanding, there will be a need for outgrower schemes to be scaled up. Second, scaling up extension services can be cost-effective in areas where farmers already know how to grow certain crops, whereas going to a new region will initially be more costly. Vegetable companies are already considering expanding to western Kenya. Third, local markets are now developing for such crops, and supermarkets are pursuing market segmentation and branding strategies around higher quality standards, different varieties, and organic or "safer" produce. There will therefore be a greater need to organize farmers and scale up extension approaches.

CONCLUSIONS

As one of the leading exporters of high-quality vegetables and fruits from Kenya, KHE has enabled over 2,500 small farmers to access European markets. KHE has an effective extension model, providing inputs, suggesting planting programs, offering technical advice, and buying back from contract farmers in cooperatives or SHGs. The key services include provision of seeds, fertilizer and CPPs (arranging providers), and planting schedules, as well as guidelines to meet GAPs and to comply with food safety and quality standards. Over the years, KHE developed a QMS under which all the on-farm activities and cash flows are recorded and compliance is well documented. KHE also has an exceptional outgrower management model, with clear division of the roles and responsibilities of agronomists from management level to field technical assistants. Both farmers and KHE benefit from the export business through adding value at different points of the vegetable value chain.

This case study exemplifies how private extension works in the horticultural export industry. It has proved effective in meeting farmers' various needs in a specific value chain. The extension model can be replicated in other high-value crops and in other geographies.

In a country with inadequate public extension services, private extension fills an important vacuum. Although only a limited number of farmers are served by such endeavors, its effectiveness and impact are remarkable. As the country advocates pluralistic and demand-driven extension services, KHE is a fine example of demand-driven approaches. To reach more and different types of farmers, both public and private actors and their collaboration are needed to address the increasing complexity of extension financing and delivery.

REFERENCES

HCDA, 2010. Horticultural crops production report. <http://www.hcda.or.ke/tech/cat_pages.php?cat_ID=73> (accessed in May 2013).
HCDA, 2013. Horticultural crops development authority. <http://www.hcda.or.ke/tech/cat_pages.php?cat_ID=73> (accessed in May 2013).
Owuor, A., 2008. The Kenya Horticultural Exporters Ltd experience of private voluntary standards. <www.agrifoodstandards.net> (accessed in April 2013).
SNV, 2012. The Beans Value Chain in Kenya. Netherlands Development Organization., August 2012.

CHAPTER 6

Private Technical Assistance Approaches in Brazil: The Case of Food Processing Company Rio de Una

Suresh Chandra Babu[1], Cristina Sette[2] and Kristin Davis[1,3]

[1]International Food Policy Research Institute (IFPRI), Washington, DC, USA
[2]Institutional Learning and Changes Initiative (ILAC), Bioversity International, Rome, Italy
[3]Global Forum for Rural Advisory Services (GFRAS), Pretoria, South Africa

Contents

Knowledge Driven Development.
DOI: http://dx.doi.org/10.1016/B978-0-12-802231-3.00006-1

INTRODUCTION

This case study documents an example of technical assistance provided by a private-sector firm in Brazil to help small-scale farmers increase their productivity and market access, thereby increasing their income. The public extension system in Brazil has undergone a major set of reforms in the past 20 years in order to meet the extension and advisory service needs of small-scale farmers. The extension and advisory needs of large-scale commercial farmers are met directly by commercial input suppliers, a structure similar to those found in developed countries.

The extension system for small-scale farmers in Brazil, however, tries to address the diverse needs for rural development activities. The system is funded by federal, state, and municipal budgets, depending on the state, under a national policy called PNATER—Política Nacional de Assistência Técnica e Extensão Rural. Some state extension agencies, like EMATER in Minas Gerais, receive large sums of federal funds, which are passed on to municipalities, while CATI-São Paulo does not receive federal funds, relying only on state budgets. Each of the state extension agencies (called EMATER in most states) employs extension agents at the municipal level and supports them with a salary. However, municipalities are expected to pay the operational costs for extension agents to reach out to the farmers in their jurisdiction. Thus the resources available to the EMATER extension agents are dependent on the municipalities in which they operate. The public extension agents are to provide comprehensive support; technical assistance to grow crops and livestock is only one of the services provided.

The state-level extension system is designed in such a way that smallholder farmer services are integrated into other rural development programs, funded by different ministries, and facilitated by EMATER agents on the ground. For example, PRONAF (Programa Nacional de Fortalecimento da Agricultura Familiar—National Program for Strengthening Family Agriculture), better known for providing credit to smallholder farmers, is implemented by the Ministry of Agrarian Development. EMATER supports the program by assessing the farmers before they can receive credit. Similarly, the PNAE program (Programa Nacional de Alimentação Escolar—National Program for School Feeding), under the Ministry of Education, transfers federal resources to states and municipalities. This enables public schools to buy at least 30% of their fresh products from smallholder farmers, thereby supporting the local agricultural system. EMATER supports the PNAE program by providing relevant information to farmers on how to access the program and on how to plan their production, including transport and commercialization of their produce. These additional tasks of the EMATER agents, combined with the already limited number of such agents, has resulted in poor reach of extension and advisory services by the public extension system in Brazil.

In order to fill this gap, private extension has emerged as an important component of the pluralistic extension system in Brazil in order to address the needs of smallholders. In the past several years, studies have shown that small-scale family farmers are responsible for most of the food produced and consumed at local and regional levels. They produce up to 80% of food consumed in many countries in Africa, Latin America, and South Asia (Nwanze, 2013). These small-scale farmers carry out their activities despite limited market access, nonexistent infrastructure, and scarce access to productive inputs, often receiving low prices for their products (Goïta et al., 2013). In the Brazilian context, family farming accounts for 70% of food production and generates 10% of the gross domestic product (Government of Brazil, 2011).

However, the relationship between output aggregators, input supplier companies, and farmers is well documented, with its advantages and disadvantages (Dogbe et al., 2012; Clark, 2012; Ferroni and Zhou, 2012; World Bank, 2012; Kirsten and Sartorius, 2010). The Brazilian case provides opportunities to explore the role of extension in other spheres as well. What role does extension play in improving access to markets? How do farmers transform their agricultural products into cash? How do farmers and rural communities access public social services? And how do farmers manage their land and natural resources so as to minimize risks, losses, and environmental damage? In order to address these and the many other challenges faced by small-scale farmers in this complex agricultural system, the different types of relationships need further analysis.

There are numerous sectors within society that make up the agricultural and food production system. These include federal and state governments that support agricultural production through provision of credit, design and implementation of policies, and development of crucial infrastructure; local governments that support the implementation of policies by providing rural extension; nongovernmental organizations (NGOs) that provide technical and logistical support; private processing companies that facilitate commercialization; and advanced research institutions that generate information and knowledge to support extensionists, to name just a few. The strategies that these numerous actors and players have used to come together to tackle poverty or food security are described well in the agricultural innovation systems literature (Coudel et al., 2013; World Bank, 2012). As in many fast-growing developing countries, the Brazilian public extension system is partly replaced by private extension organizations, particularly for commercial crops such as fruits and vegetables. Studying the nature, conduct, and performance of these entities could provide information on how their constraints and challenges might be relieved to help them provide better services to farmers.

In this case study, we look at Rio de Una, a private vegetable processing company located in the Paraná state of Brazil, to better understand the role of private actors in providing technical assistance to smallholder farmers in Brazil and to document lessons that can be learned from Rio de Una's technical assistance practices. This case study is the result of a series of interviews with Rio de Una's management, technicians, and farmers in an effort to understand the relationship between Rio de Una and farmers, the technical services provided, what changes this relationship has brought, and the linkages between these services, public policies, and the overall extension services provided in the region.

DESCRIPTION OF THE VALUE CHAIN (PRODUCTION AND MARKETING SYSTEM)

The production of vegetables to meet the demands of the newly emerging fast food and restaurant industry is the prime driver behind the commercialization and movement toward a value chain–based production system for Paraná smallholder farmers. Several private-sector output aggregators have entered this market, providing services and adding value to the produce and in the process earning a share of the additional profits that come from the organized nature of a value chain. Rio de Una is involved in a number of the value chain nodes to support the delivery of a valuable product to the market in collaboration with many actors, including farmers. Figure 6.1 illustrates the linkages between the value chain and the service providers to the farmers. We describe the actors and players in the value chain in more detail below.

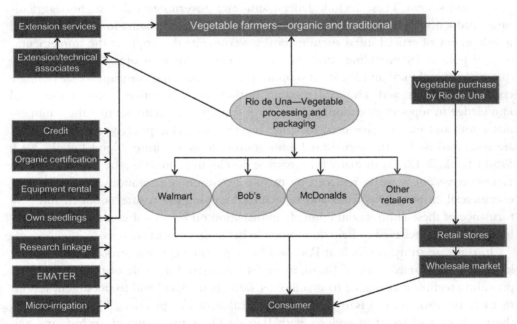

Figure 6.1 Vegetable value chain and the role of private extension in Brazil. *Source: Author's compilation.*

Rio de Una

The food processing company plays the role of the value chain organizer by connecting the farmers to the buyers of the fresh vegetables. Rio de Una (http://www.riodeuna.com.br/) is a food processing company located in the outskirts of Curitiba, the capital of Paraná, a southern state of Brazil. The company initiated its operations in 1996 as an organic processing company, during a period in Brazil when civil society organizations intensified discussions on principles of sustainability, which led to an increase of production of and search for organic products (Alves et al., 2012). However, though the company initially produced only organic products, today it also produces conventional products for a well-defined market. This choice was made due to the high cost of organic production and lack of demand in the market to absorb all the organic produce it could aggregate. Therefore, the company expanded its market and added conventional products to its portfolio to increase its profit. The total revenue of Rio de Una is roughly 50–60 million reais (US$25–30 million) per year, with a 7% profitability, which is considered high for a vegetable processing company in Brazil. About 60% of the company's profit comes from selling conventional products, such as lettuce and tomatoes, to corporate clients, as indicated in the next section. While the demand for organic vegetables is increasing, organic products in Brazil are still a novelty and it is still a modest market. Rio de Una currently processes 1,500 tons of fresh produce per month to goods ready for consumption.

Rio de Una's board of directors expressed an interest in expanding the factory to produce more conventional fresh vegetables for fast food chains, instead of organic vegetables for supermarkets. Organic products do not have the aggregated value of conventional vegetables and have much higher operating costs in terms of supervision and marketing. The directors also discussed the possibility of creating smaller versions of the Rio de Una factory in central parts of Brazil, closer to niche markets, for conventional and organic products.

Rio de Una's management has no recognized competitors in the Brazilian market as most current food processing companies lack the standards and quality control measures that supermarket chains and fast food companies require. The company's management also indicated that about 95% of all food processing companies are unable to deliver fresh processed products, since the investment needed in this area is extremely high. As an emerging economy, the current trends in Brazil include an increase in purchasing power of a majority of the population, and improved access to information about food and nutrition, all of which has led to a more demanding market for fresh foods. Prices of fresh and quality products will increase, along with increased labor costs, due to a large rural exodus to urban areas. Only capable producers will survive in the future, namely those with higher productivity and access to technology, as indicated by Ekboir (2012).

Food processing activities

The major value addition activity of the company is the processing of raw vegetables for markets and marketing outlets by improving the cleanliness and packaging for convenient use. Unprocessed vegetables are brought to Rio de Una on trucks by the farmers themselves or by third parties. The reception section collects a sample (about 2% of the total delivered) for physical and chemical analysis. A technician checks the delivery for insect infestation, condition of the vegetables, physical appearance, as well as levels of pesticide contamination and residue. If the sample is contaminated, the entire load is rejected, and the cost is covered by the farmer. Two consecutive rejections imply a breach of contract and the farmer may no longer supply vegetables to Rio de Una. These standards create a major incentive for farmers to adhere to the quality standards expected by the company. Once the vegetables are accepted, they will enter the factory to start the process of washing, selection, packaging, and storage, before being loaded into refrigerated trucks for distribution to the market. These facilities are isolated and cooled to keep from contamination and increase shelf life.

Clients

Rio de Una works with two sets of clients: those who demand conventional vegetables and those who demand organic vegetables. The conventional vegetables are sold to restaurants and large fast food chains, such as McDonalds, Subway, and Bob's, located in cities in the southern and southeastern regions. Organic vegetables are sold to large supermarket chains located in highly populated cities in southern states of Brazil (Paraná, Santa Catarina, and Rio Grande do Sul). The company tried to expand to other states where there was high demand, but due to logistical difficulties it kept its clients in the same region where the factory is located.

Farmers

A typical smallholder farmer in Paraná operates on about 10 hectares of land, about half of which is under production at any given time. For these smallholder farmers, moving from traditional production practices to organic agriculture has been a recent strategy to add value and to enter a niche market, in hopes of increasing overall income. Further, in the last 10 years most small farmers have moved away from growing cereal staple crops to growing high-value crops. The technical services provided by the public extension system have not been able to adequately cover the needs of all smallholder farmers, and the modifications that the farmers have adopted, such as high-value farming, require more frequent attention and a greater knowledge base from the extension agents.

Rio de Una has ongoing contracts with approximately 60 growers of conventional and organic vegetables, spread across four states. The process of selecting producers starts with Rio de Una identifying producers with the right capabilities to deliver the vegetables that Rio de Una commercializes. Sometimes the public extension agents—EMATERs of the state—facilitate the process of identifying potential producers through their knowledge of farmers in a particular region. The selection of farmers depends on their capacity to produce, their willingness to adapt to the standards of Rio de Una, and their geographic location. Because the vegetables Rio de Una commercializes are bounded to seasonal and climatic conditions, Rio de Una looks for producers located in diverse climate zones to ensure a year-round supply.

Farmers are responsible for the delivery of their produce to the Rio de Una factory. Some farmers have their own trucks; others pay for a private truck. The greatest distance between growers and Rio de Una is about 600 km, but the majority of farms are located within 50 km of Rio de Una's factory.

Additional markets

After supplying Rio de Una with organic and conventional vegetables, some farmers sell their surplus production at a local market, wholesale markets, or to large state-managed commercial centers called CEASAs, which exist in every state and are close to densely populated cities. Farmers reported mixed feelings about selling to the CEASAs. While it is an option for the surplus, the prices received are usually very low. On rare occasions, the market price is high and farmers make a good profit, making it more lucrative to sell to CEASAs than Rio de Una. However, these are unstable markets and farmers prefer to sell to Rio de Una to guarantee a stable income, selling only the surplus to CEASA. Occasionally, the price of a particular vegetable is so low that farmers prefer to leave the surplus on the land than take it to CEASA.

Value-added processing, on-farm processing, and packaging

Some farmers deliver pre-washed and packed vegetables to Rio de Una, adding value to their production. The installation of mini-processing factories received guidance from Rio de Una, which does the quality control on processed products in the same manner as for raw products. Farmers who have small-scale processing facilities count on family members to assist, and by doing so, increase the family's revenue. Other farmers would like to process the vegetables they produce, but lack the manpower to do so.

DESCRIPTION OF THE TECHNICAL ASSISTANCE APPROACH

Training and technical assistance provided by Rio de Una

Table 6.1 presents the various technical services provided by the company at different stages of crop production. The technical assistance provided by Rio de Una to conventional growers is less specific than what it provides to organic producers. None of the

Table 6.1 Services provided at different stages of crop production

No.	Stages of production	Extension and services provision	Contracting/integration
1.	Processing of vegetables	Company offers advice on quality of produce	Offer farmer contracts
2.	Collecting and transporting vegetables	Harvesting methods and packaging	Mobilization of farmers
3.	Quality testing for maximum prices	Quality control advice provided and supervised throughout crop season	Linking the standard of the end user and retailers to farm-level education
4.	Weeding and intercultivation	Information sharing on organic control of weeds	Mobilize farmer-to-farmer information sharing on organic control of weeds
5.	Irrigation/ fertilization	Connect to micro-irrigation companies/advice on organic production	Connect to micro-irrigation companies
6.	Pest/disease control	Provide advice/credit/ chemicals for traditional farmers	Production of input supplies and linkages to input suppliers
7.	Planting seedlings	Organic supply of seedlings	Organize linkages with seedling supply company
8.	Land preparation/ maintenance for organic agriculture	Advice on organic agriculture methods/help with certification	Organize farmers for organic agriculture mobilization

Source: Authors' compilation.

farmers producing organic products had previous experience with organic production, and this required that company staff be very involved at the beginning, with numerous training sessions and visits. The initial investment by Rio de Una was high, and therefore the selection process to choose which farmers would produce organic vegetables was conducted carefully. Rio de Una has very detailed documentation regarding what vegetables are required, either conventional or organic, accompanied with pictures, which is presented to the farmers, explaining what types of vegetables must be produced and under what quality standards.

The technicians employed by Rio de Una support many aspects of the organic production process in order to enable farmers to produce the highest quality vegetables possible. Pests and diseases are controlled by the technician, and although these services seem free, they are embedded in the purchase price. Rio de Una has four full-time technicians, and there is a part-time technician who is also employed by the state extension agency called EMATER Paraná. Rio de Una conducts periodic analysis of the water from each property, at no extra cost, to check for chemical residue.

The assigned technician develops a spreadsheet with the farmer to monitor production and related issues. The technician assists the farmers from the very beginning, from acquiring seedlings and preparing the soil, to the moment the vegetables arrive at the Rio de Una factory. The technician also informs farmers of the quantity Rio de Una will buy at each harvest, and the purchasing price.

Once farmers enter into a contract with Rio de Una, a Rio de Una technician works closely with them to assess the best production options and trains farmers on how to produce vegetables that will meet Rio de Uno standards, for either conventional production or organic. At the beginning, the technician visits are frequent, but over time the technicians' visits become less frequent as farmers develop their knowledge base on how to control pests and diseases, to manage for adverse climate conditions, to access good seedlings, or to produce according to the standards. The assistance is more intense and frequent for the organic farmers, as it is a new production system for most farmers in Brazil. The contract between the farmer and Rio de Una states that Rio de Una agrees to buy a percentage of the production (which varies from 50% to 80%) if it is in accordance with the standards, and producers agree to provide this quantity of vegetables for Rio de Una.

Farmers are classified as good, average, or poor, based on the quality of vegetables delivered. If a producer delivers substandard vegetables, the yield is rejected and he/she receives a warning. If the farmer again delivers substandard produce (classified as poor), the farmer is no longer part of the program. In terms of genetic material, Rio de Una will provide seeds and seedlings, with the cost embedded in the purchasing price. However, farmers are free to buy from a local provider or use their own, as long as it is the correct variety. Most farms are specialized, growing only one variety and/or one crop.

Farmers are not always sure about the change from conventional to organic, and a trusting relationship between the technician and farmers is key. Interviews with organic farmers showed that they were not aware of the risks and opportunities of organic farming prior to being approached by a technician from Rio de Una. They all were skeptical of the new system, and afraid of big losses and with the risks of pests and diseases. At the beginning, there were losses and a difficult learning curve, but most farmers were ultimately confident with the switch, and continue to produce organic to sell not just to Rio de Una, but also to local markets.

Some organic farmers started with only a trial plot of land to test the difficulties and advantages. Other farmers completely changed their production without experimenting, since they were convinced that the advantages of organic were greater. Other farmers experimented with organic farming, but after a few years returned to conventional production, finding it more manageable given the available resources. This speaks to the need for Rio de Una to be able to provide a wide range of services to meet the needs of different producers.

Farm-level profiles

This case study is primarily based on farm visits in the Paraná state, where Rio de Una has its major procurement operation. The company was highly cooperative in connecting the researchers with farmers with various backgrounds and growing different crops. This helped to get a broad sense of the farming operations under the technical assistance of Rio de Una.

The social system of rural Brazil is mainly of a patriarchal structure, where the father usually owns the land, and his wife and children provide labor and can participate in the management and decisions of the land. All four visits described below follow this model.

Organic farm 1

The first farm visited, which made the transition to organic farming 10 years ago, has 38 hectares of land supporting entirely or partially three families: father (owner of the land) and mother; a daughter and her family; and a son and his family (only partially, as his wife is a strawberry farmer on a different property). The farm produces primarily vegetables—cabbage, broccoli, and lettuce—the most common vegetables under the contract of the company. The farm is operated by family labor. The daughter and her husband who live on the farm provide major labor and management support. The son of the farmer, who lives about 5 miles from the farm, provides additional labor on a daily basis. In addition, a rural worker is hired to support the farm's labor when needed. The farm has a tractor that is about 10 years old, and there are drip irrigation facilities on the farm. The support given by Rio de Una in terms of transformation from a traditional farm to an organic farm and the associated benefits are appreciated by the family. The agronomist from the company is held in high esteem by the family and is relied upon for other matters related to solving family financial issues. The main advantage pointed out by the farmers was that they now plant and harvest every 15 days, all year long, instead of harvesting over irregular periods of time. With that, they have managed to paid old debts and stabilize their finances. In addition to vegetables, the mother keeps cows for production of cheese, which is sold in town for additional income. The vegetable waste and surplus are converted into cattle feed. The profit made in the first years was reinvested in the farm, buying inputs and machinery.

The farm operates under the Rio de Una certification of organic production, but the farmers have requested from the municipality to have their own certification, to be able to sell their certified products to other buyers. Based on the expansion of organic producers in the region, the municipality facilitated the formation of an association of 14 organic producers, and there are plans to open an organic shop in the city hall; however, most associates produce small quantities of organic vegetables and fruit, not in large scale, like the farmers interviewed. The aim of the association is to open new markets.

Besides learning from the Rio de Una agronomist and printed material provided by Rio de Una, the farmers exchange information and search for other sources of information (mainly TV programs and printed material provided by the municipality) to control pests in their production. At the beginning, Rio de Una organized field days, where farmers could learn from each other. Today there is no such initiative, except during the initial steps of forming an association.

The farm regularly supplies vegetables under the contract to the company and sells its surplus to a regular market and/or to another purchaser in the region, who was a contract farmer for Rio de Una but has since taken up aggregating commodities. However, this local purchaser, a Rio de Una farmer turned market aggregator, does not provide any technical advice. The family purchases seedlings from a designated agent who is certified by Rio de Una. Organic farming is still an experimental and innovative approach for the family farm and for the agronomist, who shares ideas and techniques learned from other farmers. Other major concerns among farmers are the replacement of labor, having enough rural workers to continue or expand production, and the division of the land among five children in case the father passes away. In the eventual division, the size of the plot for each child would not allow sufficient income to keep a family.

Organic farm 2

The second farm visited is 5 hectares of prime land which was converted from traditional farming to organic farming. The change has been positive for the family as the organic vegetables fetch more money for less cost. The farmer plants lettuce throughout the year and uses a drip irrigation system. He has a tractor and a medium-sized truck, both purchased with the extra income that the organic farming brought in the last 4 years. The truck was financed through the PRONAF rural credit program. The farmer relies on family labor; his two daughters, son-in-law, and wife all work with him on the farm on a daily basis. The company provides technical assistance and an assured market and prices that are known well in advance. He also takes the surplus produce to other marketing agents and to the local market. The price in these markets fluctuates greatly, and the farmer considers this as additional income that varies according to the market.

With technical assistance from Rio de Una, the farmer installed a mini-processing factory, where vegetables like cauliflower are selected, washed, and packed in individual trays. The factory is operated by the farmer's daughter. Rio de Una stimulates farmers to invest in such mini-processing factories, as there are benefits for both parties.

Conventional farm 1

The third farm visited is 12 hectares, with 4 hectares dedicated to production of lettuce, cabbage, pumpkins, and other products that are in demand. The farm is managed by two business partners, with two additional rural workers during harvest.

The lettuce production is mainly designated to Rio de Una, while cabbage is sold in the regional wholesale market called CEASA. The farmer tried organic farming for a few years, but decided to return to conventional production after disagreements with Rio de Una management and because of a lack of alternative buyers in the region. The technical assistance received comes from Rio de Una and also from agronomists from input supplier companies. This assistance is not paid for by the farmer. EMATER is not active in the region, and farmers rarely receive a visit from EMATER. But EMATER assisted with the request for rural credit from PRONAF. When production is high, it is an advantage for the farmer to sell to Rio de Una, because products are commercialized by weight. If the weight of the vegetables is low, profit is not made, but matches the costs.

Conventional farm 2

The fourth farmer is a conventional farmer who cultivates about 6 hectares of land, out of a total of 10 hectares, which includes protected areas as well. The farm is managed by a family of three: father, mother, and a young son, who helps after school. An older daughter is employed in town, after gaining a degree, and has her own family. An older son, who was active on the farm, died of cancer few years ago. The young son plans to apply for university as well, and does not intend to be a farmer.

The farm cultivates lettuce and endives, both under the technical guidance of Rio de Una. The lettuce is designated exclusively for the fast food chain McDonalds, through Rio de Una, and the endives, due to an excess on offer, are not purchased by Rio de Una anymore but sold to markets in São Paulo state or CEASA. During winter all the production is for Rio de Una, but during summer, when harvest is abundant, the farm diversifies crops and markets. The agreement with Rio de Una is considered positive, as being a more steady market than CEASA. The farmer spent 26 years producing vegetables and selling mainly to CEASA. Previous to Rio de Una, the farmer had a short-term experience providing horticulture to Nutrimental, a food processing company in Paraná. Nutrimental had an approach similar to Rio de Una's in terms of partnership and technical assistance. The partnership with Rio de Una is perceived as profitable, but a lack of technology and infrastructure (such as greenhouses) prevents the farm for being more lucrative. There is enough land to produce more, but not enough labor. A member of the extended family helps during harvest.

Before the relationship with Rio de Una, technical assistance was provided by input supply companies. EMATER has not provided technical assistance to the farm. Seedlings are purchased by the farmer from regional companies. The farmer had benefited from federal credit (PRONAF) and acquired a new tractor, to be paid in 10 years (interest of 2% p.a.). In this case EMATER did not facilitate the process, but rather the machinery dealer, who charged a commission to help the farmer to fill out the necessary documents.

For 2 years, the farmer was a treasurer for the producers association of Paraná, and visited CEASA Curitiba many times to discuss internal administrative issues of CEASA, not to negotiate prices and conditions on behalf of producers. CEASA refers to producers as partners, but farmers feel they are suppliers, not real partners. Functional associations of producers are rare, as association directors are perceived as dishonest, incompetent, individualists, and highly politically oriented.

ANALYSIS OF PRIVATE TECHNICAL ASSISTANCE

The proposed framework for analyzing technical assistance in Brazil considers six variables: relevance, efficiency, effectiveness, equity, sustainability, and impact. Based on the interviews and observations carried out with the different actors, the researchers provide an analysis according to these variables.

The private sector approach to technical assistance in the case of Rio de Una derives from its need to have higher-quality produce according to the specifications required by their clients. The relevance of technical assistance in both the organic and conventional cropping systems comes from the specifications set by the final purchasing companies and regulatory agencies, especially in the case of organic products. For example, the quality requirements of these companies require the agronomists to provide sufficient technical support to meet these standards. Organic agriculture, however, requires teaching the farmer a new system of farming. This means the agronomist must align training materials to the company's goals as well as the information and training needs of the individual farmer. This system of technical support may not be relevant for conventional farmers who are not part of the contract (because different crops may be grown to different standards). Even contract farmers sometimes need information beyond crop growing assistance, which a private company is not willing to give, as it is beyond its mandate and not financially feasible for them to do so. For these purposes farmers depend on the EMATER agents, who are spread thin in the municipalities covered by the company.

The technical assistance offered by Rio de Una is considered efficient since it is focused on a set of farmers whose needs for crop production are well defined, and the goals they need to achieve are stipulated in the contract between the farmer and the company. The agronomist ensures that the farmers' production constraints are addressed in a timely manner, by adhering to a streamlined schedule of contacts and visits to the farms. The organic farmers are visited by the supervising inspectors. The company also plays the role of middleman between the input suppliers and the contract farmers. This makes the system of technical assistance provided by the company highly efficient from both the farmer's and the company's perspective.

The effectiveness of the technical assistance system can be assessed according to the extent to which the goals of the company are achieved and/or the extent to which the farmers' technical assistance needs are met by the technical agents. From the company's perspective, agents are able to reach out to the farmers throughout the season and help them with their technical and input needs. As mentioned earlier, the farmers' needs often extend beyond the technical assistance provided by Rio de Una.

The private technical assistance system could address the vast gap left by the public extension system and the gap in technical assistance, which large farmers can access by hiring specialists to solve their problems. These large farmers are also linked to the national research system—EMPRAPA—while the smallholder farmers served by the private company depend on the state extension agencies. However, due to low coverage of the EMATER system in many states, the private company plays the role of linking smallholder farmers to new innovations and technologies. This increases the level of equity in the delivery of the extension system in Brazil.

The sustainability of the extension system is ensured as long as the farmer fulfills his contractual obligations and complies with the quantity and quality requirements. However, the extension needs of the farmers who are not contracted with the company must be met by the publicly funded extension system. Most farmers outside the company's contractual arrangements grow vegetables using conventional methods. Since these farmers are not fully served by the public extension system, the sustainability of this production system is in question.

The impact of private assistance on the contract farmers in Paraná state can be seen in several ways. First, the technical agents of the private company have been successful in transforming some farmers from conventional agriculture to organic agriculture to meet the growing demand for organic products. Second, the production contracts and the associated technical advice to meet quality standards have helped to assure a market for the farmers' products. Third, due to the involvement of the private company's agents in guiding and supervising the farmers in following the rules and regulations for organic certification, the process of labeling their produce as organic has improved.

Perspectives of the producers

Researchers visited four producers providing organic and conventional vegetables to Rio de Una. The technical assistance provided by Rio de Una is specific to the products they commercialize; they do not cover all crops or livestock raised in the region. Farmers indicated that the learning curve was difficult and stressful, but they were satisfied with the services they received because of the results that were generated. Farmers improved their livelihoods and gained a more secure income; although it varies from month to month, some continuity (and overall increases) was observed. Farmers who produce organic vegetables are more satisfied with the decrease of contamination risks involved in managing and applying chemical products.

Farmers made improvements to their properties with the additional income they received; but while most of them have the land and knowledge to increase yields, they lack manpower, particularly farmers producing organic lettuce, or the infrastructure to do so.

The challenges faced by farmers are also related to the price received for the products sold to Rio de Una. The company agrees to buy a certain percentage of the production during the harvest, at an agreed price, but the price varies based on the market conditions. Farmers claim they receive a low price for their products, but believe selling to Rio de Una is still a good option for them, since they have a relatively stable income and receive specific technical assistance.

Farmers also showed some concern about future activities, especially about who will be taking over the farm in the future, since most children have no interest in farming, aiming to achieve higher education and city jobs.

Perspectives of technicians

Researchers spent some time with Rio de Una's senior technician to understand the approach taken by the company. A former rural extension agent from EMATER, together with three other technicians, she provides technical assistance to about 60 farms. She believes the relationship between technicians and farmers is good, and smallholder farmers who are part of the program are better off than nonparticipants. However, she emphasizes that not all farmers are qualified to enter a private contractual arrangement such as that organized by Rio de Una, whose standards are high. Several farmers' contracts were cancelled because they could not provide the products at the quality required. The technical coordinator emphasized the importance of technical assistance for the success of the program, where farmers, especially those in organic farming, have to learn new techniques to meet national organic standards. Personal commitment and belief in the business model of Rio de Una was key to the success of the program in its early years. Farmers have acquired the basic knowledge they need and minimal technical assistance is needed currently to continue to support those who have been in the program for some time. The technicians noticed improvements in the livelihood of the farmers due to increased income generation.

The company started a partnership with EMATER Paraná and values that connection. However, researchers found that although Rio de Una is establishing a relationship with EMATER, the partnership is still modest and could be enhanced. There are several issues that hinder the development of a better relationship between a processing company and public institutions, such as bureaucracy or the shortage of technicians from EMATER.

Perspectives of company management

Rio de Una's management is proud of the business model developed, stating that "the model has to work; otherwise Rio de Una would not exist." The company's director believes Rio de Una has a good relationship with farmers and that both parties benefit from the collaboration. Farmers learn new techniques and have stable revenue, and Rio de Una supplies its market and generates profit. Companies like Rio de Una are able to reach farmers better than most public institutions. Future business opportunities are greater than ever with the changes in consumer habits. The Brazilian population is eating out more often, and has less time to prepare meals at home.

Perspectives of the community at large

When discussing the Rio de Una's activities and approach, rural extension academics see the approach as positive, providing benefits to farmers who are part of the program. Extensionists feel uncomfortable that private food processing companies, like Rio de Una, provide rural extension because of the mismatch of incentives. According to extensionists, private companies provide specific technical assistance, which aims to improve the product and the supply, not necessarily the livelihood of the rural family or the environment. Most companies are not interested in charity or the well-being of farmers, concentrating on their loyalty and capacity to deliver high-quality products. Social responsibility is a new concept in Brazilian society, and it will take some time for private companies to include this concept in their business models.

COMPARISON WITH PUBLIC EXTENSION AND OTHER PROVIDERS

The public extension system in Brazil has undergone a major set of reforms in order to assist smallholder farmers. Extension is generally viewed as technical assistance that goes beyond simple technology transfer. As in other developing countries, extension agents are called upon to assist with related activities. In Brazil, the EMATER agents are commonly called upon to assist farmers in the preparation of applications for other government programs that help farmers. PRONAF, the federal program to provide medium- and long-term loans to build farm capital are a classic example. However, even with this type of assistance, due to the need to provide a large number of farmers with technical assistance, EMATER agents only respond to farmers who get in touch with them.

EMBRAPA—the Brazilian national research system for agriculture—is often referred to as a successful model for promoting new technologies that are relevant for farmers. However, the National Policy for Technical Assistance and Rural Extension (PNATER) and the need for reform in extension for smallholder farmers clearly show that EMBRAPA mainly addresses the needs of the medium- and large-scale farmers and has not been relevant for the smallholder farmers PNATER tries to reach. The policy reforms in extension have been a result of this neglect. Recently the government announced that it will bring together EMBRAPA and PNATER through an agency called ANATER, created to effectively use EMBRAPA's technologies.

To understand the role of EMATER in the extension services at the state and municipal levels, key leaders of EMATERs in three states were invited to Brasilia for a one-day workshop. In addition, extension officials from the Ministry of Agrarian

Development were consulted to understand the historical background of the public extension reforms. The highlights of the discussion are discussed below.

From the discussions, it is clear that the public extension system in the form of EMATER is not as responsive as the private extension provided by companies such as Rio de Una. What emerged from the discussion with the private extension agents is a feeling that EMATER should play a broader role and a form of public–private partnership should be developed in the areas where private extension is currently active. Engaging farmers and identifying needs should still be the role of EMATER. However, EMATER could play a facilitating role in preparing farmers to connect to the private sector and can play a monitoring role so that the private sector does not become exploitative in their operations.

The case study farmers who are supplying the vegetables to Rio de Una under contract expect services beyond technical assistance alone. The private extension agents are, to some extent, to get involved in these additional supporting roles to maintain business relationships with the contract farmers. However, the manager of the company indicated that he is not willing for his technical agents to go beyond the technical assistance role, as this would involve mission creep as far as the company is concerned.

The Rio de Una example has inspired other private-sector agents to buy produce from farmers in the Rio de Una contract region as well. While these agents are interested in providing an outlet for the farmers' surplus produce, they are not particularly inclined to offer extension the way Rio de Una currently does. This may be considered a free ride, but little can be done about this, and farmers cannot be prevented from supplying commodities elsewhere. Rio de Una currently turns a blind eye to this practice in order to maintain good relations with their contract farmers. Thus there is no viable alternative company in the region to replace Rio de Una and its extension services.

In summary, the technical service provided by private companies such as Rio de Una are distinct from public extension services provided by public agencies (e.g., EMATERs) or other agencies with public funds (e.g., NGOs). Most public extension services include technical assistance, but go beyond assisting farmers to increase production and productivity. Public extension services aim to improve the livelihoods of farmers and rural communities by exchanging knowledge and supporting efforts for better health, sanitation, education, income, natural resources management, and preservation of cultural practices and beliefs.

UNIQUENESS OF THE MODEL AND VALUE ADDITION

Rio de Una provides specific technical assistance to improve the quality of the products its company commercializes. Farmers are well trained in how to produce conventional and organic products according to the requirements of the company and regulatory agencies. It is a win–win situation, in which it is in the best interest of the company that farmers be engaged and produce better vegetables, and farmers have a market for their products, generating a stable income. The model engages farmers in a profitable business, and the model has proved attractive to young farmers as an alternative to rural exodus.

LESSONS FOR REPLICATION AND SCALING UP

The experience of Rio de Una identifies gaps in rural extension that could be filled by output aggregating companies. This will require improving the linkages and coordination among all the actors in the value chains, including the market for produce, the aggregating company, extension agents, and farmers. The company has been able to identify gaps in technical assistance and fill those gaps. This model is replicable and scalable if the market for processed vegetables is identified.

A major benefit for the farmers who are under contract with Rio de Una is the assured market and the near elimination of the uncertainty in market prices. The technical agent is also able to advise on the type of crops grown depending on the natural resource conditions and agroecology of the region. This assured market and the extension services that go with it have reduced dependency on the public extension system and marketing agents who tend to exploit the farmers, using fluctuations in demand to their advantage.

The farmers under contract to Rio de Una get intensive technical assistance to deal with technical, regulatory, and financial issues throughout the crop production, postharvest, and processing stages. This interaction at the beginning of the contract helps the technical agents to gain the trust of the farmers. That trust continues throughout the contract, and the agents are present to support the farmers at all times. This has helped the agents to introduce organic farming methods and develop a specialty market for the commodities, adding further value to the commodities.

As a result of increased productivity and the stability of the income from their farms, some farmers have increased their assets and invested more in the form of pickup trucks and tractors, which further helps to increase the productivity and efficiency of their operations. The assured income flow from the farm has also helped farmers to plan their farming and investment in resource management better than before. Thus the extension-mediated income increase is seen as a binding factor for the farmers to renew their contracts with the company and continue to be part of the value chain.

The farmer–technical agent interactions have also resulted in some farmers innovating on their own farms. For example, by using locally available materials to protect their crops from pests and diseases, the technical agents provide opportunities for inclusive innovation that involves farmers. In addition, agents are able to gauge the farmers' interest in various approaches and guide them to test new practices on a few farms before they are shared with other contract farmers.

Demand for select products, such as organic products, has been effectively captured by the company, which then enters contracts with farmers for product aggregation. However, the company was not able to fully implement this transformation as the demand in the market was not high enough to make it profitable. Thus the company operates in a differentiated mode of dealing in both traditional and organic markets and adapts its technical advisory strategy according to the information needs of the farmers. This flexibility helps the company in two ways: by helping the company to meet its demand in the differentiated markets and by expanding the organic methods of farming in a phased manner as the demand for organic products expands. The company has also built its own technical capacity to help farmers with farming, regulatory, and marketing practices to meet the standards of the current market.

Although the technical assistance provided by the private company has been successful, the company operates in isolation, dealing directly with input suppliers and certifiers of organic systems. They have little if any interaction with the public research systems. This is partly due to the high level of bureaucracy in the public system, which does not allow opportunities for private companies to benefit from public research and extension systems.

CONCLUSIONS

Improving the livelihoods of smallholder farmers is a complex issue. No single provider of services, either public or private, is able to address in isolation all of the social, economic, environmental, and physical needs of farmers or of rural communities as a whole. Nor is any able to support food security and the sustainable use of natural resources in rural and urban areas.

Public rural extension plays a role in supporting the most vulnerable rural groups, and private companies play a role in supporting farmers' activities and offering the opportunity for a more reliable income. The challenge is to develop long-lasting partnerships that can make better use of the human and financial resources available in different types of organizations, under a coherent framework or national policy.

The business model of Rio de Una is characterized by its relationship with farmers with certain capabilities to produce what is required, when it is required. The characteristics of these farmers are specific, and do not reflect the abilities of all smallholder farmers. In other words, the model works for innovative farmers, and those who do not yet have the minimum capacity to work with the private sector should be assisted by public programs, which should be under the framework of national policy (PNATER) and investments made by federal and state governments.

The farmers' main concern is with income, and although public institutions can help farmers to increase production and productivity, they are unable to assist farmers with the commercialization of their products to improve the stability of their incomes. Isolated initiatives can be seen, such as food fairs and women's handicraft markets, but these are modest activities compared with the number of smallholder farms that exist in Brazil, and increasing rural poverty and rural exodus. Food processing companies, such as Rio de Una, provide a viable alternative to public extension systems for family farms that meet the minimum level of agricultural production required to undertake a contract with a private-sector company.

REFERENCES

Alves, A., Santos, A., Azevedo, R., 2012. Agricultura orgânica no Brasil: sua trajetória para a certificação compulsória. Revista Brasileira de Agroecologia 7 (2), 19–27.

Clark, J., 2012. The Effectiveness and Sustainability of the Input Supplier Model. Working Paper 9. Agricultural Learning and Impacts Network (ALINe), Brighton.

Coudel, E., Devautour, H., Soulard, C.T., Faure, G., Hubert, B. (Eds.), 2013. Renewing Innovation Systems in Agriculture and Food: How to Go Towards More sustainability? Wageningen Academic Publishers, Wageningen, the Netherlands.

Dogbe, W., Sogbedji, J., Mando, A., Buah, S.S.J., Nutsugah, S.K., Kanton, R.A.L., 2012. Partnership for improved access to agro-inputs and technology: some experiences from the emergency rice initiative project in Ghana. Afr. J. Agric. Res. 7 (34), 4790–4802.

Ekboir, J., 2012. Facilitating smallholders' access to modern marketing chains. In: Agricultural Innovation Systems: An Investment Sourcebook. World Bank Publications, Washington, DC, pp. 52–58.

Ferroni, M., Zhou, Y., 2012. Achievements and challenges in agricultural extension in India. Global J. Emerg. Mark. Econ. 4 (3), 319–346.

Goïta, M., Amougou, P.A., Mecheo, S., 2013. Family Farmers for Sustainable Food Systems: A Synthesis of Reports by African Farmers' Regional Networks on Models of Food Production, Consumption and Markets. EuropAfrica Consortium, Rome, Italy, p. 43.

Government of Brazil, 2011. Key facts & figures: agrarian development and food policy in Brazil. 37th Session of the UN Food and Agriculture Organization (FAO) Conference. Rome, Italy. <http://www.brasil.gov.br/para/press/press-releases/june-1/brazil-to-highlight-south-south-cooperation-initiatives-at-fao-side-event/files/fact-sheet-agrarian-development-and-food-policy-22jun2011.pdf>.

Kirsten, J., Sartorius, K., 2010. Linking agribusiness and small-scale farmers in developing countries: is there a new role for contract farming? Dev. South. Afr. 19 (4), 503–529.

Nwanze, K.F., 2013. Viewpoint: Smallholders Can Feed the World. International Fund for Agricultural Development (IFAD), Rome.

World Bank, 2012. Agricultural Innovation Systems: An Investment Sourcebook. World Bank Publications, Washington, DC.

CHAPTER 7

Syngenta Frijol Nica Program: Supporting Nicaraguan Bean Growers

Yuan Zhou
Syngenta Foundation for Sustainable Agriculture, Basel, Switzerland

Contents

INTRODUCTION

Nicaragua's agricultural sector contributed about 19% of the country's GDP in 2012 and employed 32% of the economically active population (World Bank, 2013). Raw coffee, beef, sugar, peanuts, rice, poultry, and horticultural crops are among the main agricultural products. In recent years, several factors have hampered agricultural productivity in Nicaragua, including farmers' limited access to quality inputs and credit, inadequate infrastructure and marketing services, and volatile weather.

Knowledge Driven Development.
DOI: http://dx.doi.org/10.1016/B978-0-12-802231-3.00007-3

Nicaragua produces about 70% of the red beans consumed in Central America; they are a key income-generating and food security crop for rural communities. There are approximately 250,000 small growers of beans, each producing on an average of one hectare of beans. Prior to 2000, beans were only grown for subsistence farming. Farmers focused production efforts on household consumption and sold to the local market only when they had a surplus. The industry has since gradually shifted to cash cropping, selling dry beans both nationally and internationally. Dry bean exports increased from US$7.5 million in 2000 to over US$60 million in 2009 due to demand for red beans in Central America and the Central American population in the United States.

The majority of smallholder farmers have not been recipients of agricultural extension services in the past. Public extension in Nicaragua has only limited coverage due to financial and human capital constraints. Prior to 2006, most bean farmers did not use high-tech agricultural inputs, and had low yields and limited access to loans and technical assistance. In response to these limitations, Syngenta established the Frijol Nica (Nicaraguan Bean) program to provide growers with technology, access to credit, and technical support so they could increase their income. This program exemplifies how the private sector can fill a gap in agricultural extension and innovate in the area of service delivery.

Information in this case study is largely drawn from a series of interviews and discussions with the company, its various partners and farmers, and with public extension workers. This chapter focuses on the dry bean value chain in Nicaragua and describes the extension of the Frijol Nica program. The relevance, efficiency, sustainability, and impact of the private extension program will be discussed, followed by lessons for replication and scaling up, and concluding remarks.

THE DRY BEAN VALUE CHAIN

In Nicaragua, dry beans are produced predominantly by small-scale growers, located in the regions of Jinotega, Matagalpa, Boaco, Nueva Guinea, Carazo, and Estelí. Farmers cultivate red and black beans to be sold to traders in the market. A number of cooperatives work together with traders to collect dried beans from farmers. Traders then sell the beans nationally and internationally (currently this is mostly black beans).

After the 2001 worldwide coffee crisis, Nicaragua needed to focus on other agricultural commodities for economic stability. Agricultural input companies, such as Rappaccioli McGregor, S.A. (RAMAC) began to look for alternatives and decided to invest in the bean industry. In 2003, the government started the "Libra por Libra" program to encourage the use of certified seeds to improve production. The government purchased certified bean seeds from seed producers and then distributed them to grain farmers at a discounted rate. The program continued for 3 years and served thousands of bean growers. Even after the official end of the program, the government continued to purchase certified seeds until 2008. Before 2008, the government stayed away from price control in the bean industry. But due to opportunistic trading activities conducted by El Salvador, the state-owned company ENABAS (Empresa Nicaraguense de Alimentos Basicos) started to buy beans and store them to absorb negative effects from the price speculations of 2008. When bean prices increased, ENABAS would sell the beans to increase supply and stabilize prices. Given strong bean production and attractive prices in 2008, many farmers planted a larger area of beans than before. This resulted in an oversupply, causing prices to drop in early 2009. When the red bean price remained low, the government encouraged the production of black beans, offering a higher fixed price. Motivated by the higher price, some farmers switched to black beans, thus increasing the proportion of black bean production. Nicaragua has been exporting black beans to Venezuela since early 2007 under ALBA, an alliance of nine nations in Latin America sponsored by Venezuela. Only four companies in Nicaragua are permitted to export: Esperanza Coop, FENIAGRO, Nicaraocoop, and FENACOOP.

Key trends of dry bean production

In Nicaragua, bean consumption is extremely important culturally, and beans are considered a staple food. Most farmers cultivating red beans produce primarily for their own consumption and sell their surplus locally. They normally farm beans with other crops such as maize, vegetables, or some fruit trees. In contrast, black beans and other type of beans are mainly produced for export markets. At present, red beans account for about 90% of total bean production, whereas black beans account for 10%.

According to data from the FAO, the average area of production and yields vary from year to year; 2011 saw the highest production in the past few years (Table 7.1). Reported yield levels were considerably lower than other observations and reports from Nicaragua. According to the Union de Productores Agropecuarios de Nicaragaua (UPANIC), in 2011/2012 the average yield of various types of beans was about 1.7 tons per hectare nationally. Export values showed a declining trend between 2008 and 2011, but picked up in 2012, yielding an export value of US$51.8 million.

There are three seasons for dry beans in Nicaragua: primera, postrera, and apante. Primera, the smallest harvest of the year, is planted in May and harvested in August in the dry western and northern zones. Postrera is planted in the same regions in September and harvested in December, accounting for about a third of annual production. The

Table 7.1 Dry bean production in Nicaragua

	2008	2009	2010	2011
Area harvested (ha)	239,165	248,592	216,490	303,808
Production (tons)	176,655	213,464	150,837	234,163
Yield (kg/ha)	739	859	697	771
Export value (million US$)[a]	79.8	61.5	59.4	30.2

[a]Banco Central Nicaragua, Estadisticas economicas anuales (2011).
Source: FAOSTAT.

majority of annual production comes from apante, which is planted in the northern regions and along the Atlantic Coast in December and harvested in March. The third season is particularly important because of good seasonal rainfall. In some regions beans can be planted only in the first two seasons, whereas in Sacacli in the Jinotega region beans can be planted during all three seasons.

The value chain partners of the bean sector

The dry bean value chain in Nicaragua begins with input suppliers, who distribute inputs to cooperatives or independent growers. In the case of the Frijol Nica program, the agricultural inputs are first selected and then recommended to growers in a technical sheet provided by RAMAC. Esperanza Coop acts as an intermediary and a guarantor for district cooperatives or producers to get these inputs on credit. The district cooperatives are responsible for distributing the inputs to the bean producers. From the technical sheet provided, producers know what amounts of fertilizer and pesticides to apply and when. Figure 7.1 illustrates the key partners in the value chain.

After harvest, producers can sell their beans to various traders. Red beans can be sold directly to wholesalers, national traders, or cooperatives, which will sell to other participants in the chain (Figure 7.1). Black bean farmers can sell their produce to their cooperatives or one of the export traders, such as Esperanza Coop, which then sell to importers in Venezuela and other markets such as the United States and Costa Rica. Independent producers typically sell directly to export traders. Traders usually have warehouses to store dried beans and sell when prices increase.

Esperanza Coop is one of the national traders in black beans. Its 330 members produce beans and other crops. Esperanza acts as an independent company that cooperates with producers and other cooperatives to obtain credit, improve bean cultivation, and export black beans. The cooperative works through partnerships with RAMAC,

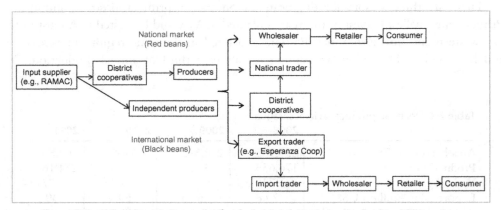

Figure 7.1 Key partners in dry bean value chain in Nicaragua. *Source: Author's compilation.*

independent producers, and district cooperatives. It then exports black beans to Venezuela under the ALBA trade agreement. In addition, Esperanza Coop distributes certified and/or registered seeds. These seeds are improved varieties developed by INTA (Nicaraguan Institute of Agricultural Technology). Esperanza Coop procures the seeds from seed producers and then sells them to cooperatives and farmers. The commercial varieties include "INTA Cárdenas" (black beans) and "INTA Rojo" (red beans). The Nicaraguan Agriculture and Forestry Ministry (MAGFOR) regulates seed production and seed certification.

EXTENSION METHODS OF THE FRIJOL NICA PROGRAM

Syngenta describes itself as a leading agribusiness committed to sustainable agriculture through innovative research and technology. Headquartered in Basel, Switzerland, it has more than 27,000 employees worldwide dedicated to its purpose: "bringing plant potential to life." It offers two main types of products: seeds and crop protection.

The Frijol Nica program was initiated to tackle the challenges faced by bean growers in Nicaragua. Its main objective is to provide an integrated package of high-level technological solutions, technical assistance, support in market linkages, and education for producers and their families.

Extension model

The Frijol Nica program was initiated by Syngenta and a strategic partner, Rappaccioli McGregor, S.A. (RAMAC) in Nicaragua in 2006. It aims to increase bean productivity and profitability to farmers, thereby increasing their livelihood. The partnership dates back to 2001, when coffee prices collapsed and RAMAC decided to invest in the bean industry. Together with Syngenta, RAMAC started field trials in bean production, testing how yields changed with the application of various levels of fertilizers, chemicals, and other inputs. Before that, beans were largely considered to be a subsistence crop; input suppliers, government, banks, and NGOs showed little interest in supporting bean producers. The pilot project run by Syngenta and RAMAC started with 28 hectares and quickly increased. The goal evolved from field testing into building a functional bean supply chain. Since then, Syngenta and RAMAC have invested around a half million dollars in training, talks, forums, and input supply to support the program and to reach as many farmers as possible. In 2008, producers were pleased with the technical support and yield increases but found market channels to be bottlenecked. Therefore, Syngenta and RAMAC created an alliance with Esperanza Coop to help them with links to market.

The Frijol Nica program is currently working with 6,500 bean growers, located in the major producing regions of Jinotega, Matagalpa, Estelí, Boaco, and Nueva Segovia. A small group of producers are situated in the South Atlantic region, but they are less well organized and have lower yields. Most farmers in the program belong to cooperatives, with the exception of a small number of independent producers.

Key partners and their roles in the program

As the originator of the program, Syngenta plays a supporting role in program implementation. The local distributor, RAMAC, is Syngenta's only direct partner. It is a company with extensive experience in the agricultural input industry, distributing products from many different manufacturers. RAMAC is in charge of implementing the program, contacting the producers, dealing with cooperatives, and searching for credit sources for the cooperatives, which then extend credit to producers. With guidance from Syngenta, RAMAC provides technical support at the farm level, and seeks new business opportunities and linkages with other organizations.

Within Syngenta, several people are involved in managing the program. At the lowest level, there is one person responsible for beans and other crops, including rice, maize, peanuts, and other cash crops. This person, together with a sales coordinator, oversees the program and tries to solve any issues with distributors, business partners, or producers. The sales coordinator reports to Syngenta Nicaragua's general manager. The company seeks alliances with cooperatives, associations, and NGOs, aiming to reach and influence more than 200,000 bean producers. Reaching so many farmers is ambitious and requires strong partnerships.

Esperanza Coop is the most important trader for the Frijol Nica program. The cooperative plays a comprehensive role: It produces beans (both seeds and grain), provides technical and financial support to producers, procures beans in different regions, and acts as a trader in international markets. Esperanza Coop is one of the few traders permitted to export black beans to Venezuela under ALBA. The cooperative distributes certified seeds that it itself has produced. It also acts as an intermediary between input suppliers such as RAMAC and the producers. It is in charge of drying the beans at a processing plant in Sebaco, where beans are prepared for export. In addition, the cooperative has its own certified weighing scales to prevent exploitation of farmers in the collection centers. It also runs quality control, in particular in terms of humidity levels, in order to maintain export quality.

Knowledge and technology transfer

The Frijol Nica program has used various methods and strategies to provide extension services to bean growers. Most extension activities include training and field visits, through which farmers acquire knowledge about the best agricultural practices to prevent pest infestations and disease and to improve yields. Technical training is offered three times a year, before the sowing of each planting cycle. Through training farmers learn techniques such as seed spacing, fertilization, weed control, and pest and disease control. Field visits occur more frequently during the growing phase, from sowing to harvesting. Farmers usually need most support and field-level advice between April and January.

In Syngenta, there are two types of field advisors: the CAS (Agricultural Counselor) who looks after large-scale producers (over 100 hectares), and the RAS (Agricultural Representative) for small-scale growers. Both are considered technical sales representatives who also provide support to distributors. Under the Frijol Nica program, there are several RAS who are responsible for meeting farmers' technology and knowledge needs. Each is in charge of about 1,500 producers registered in the program. In 1 week, the field advisors typically visit 80 farmers who are located in close proximity. Most requests for assistance concern issues related to weeds, pests, and disease control. In addition, Syngenta has field promoters who visit farms and inform producers about the program and other services provided by Syngenta. They also work with distributors.

Field advisors from the distributor RAMAC are usually first trained by field advisors from Syngenta, and they in turn train field technicians in different cooperatives. The training covers topics ranging from the use of certified seeds to post-harvest handling. Figure 7.2 shows the flow of training and knowledge transfer. Field technicians can cover the immediate needs of farmers in cooperatives. The rationale for having only a few field advisors from Syngenta and RAMAC involved in the program is that producers can also act as knowledge transfer agents. Additional help comes from Syngenta field promoters who support field advisors in assisting farmers with various needs related to the crop.

Syngenta and RAMAC work together, each bearing half the cost of some of the activities related to input supply and technical support. The cost of training activities

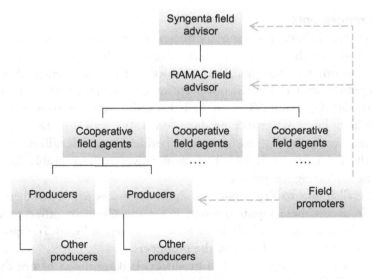

Figure 7.2 Flow of information and knowledge in Frijol Nica program. *Source: Author's compilation.*

for promoting the Frijol Nica program is entirely covered by Syngenta. This promotion now includes the campaign "Mochileros en campo," in which field promoters introduce the program to producers and their cooperatives, show the results, and encourage producers to use new technologies.

The Frijol Nica program uses the alliance between RAMAC and Esperanza Coop to train field advisors from various cooperatives that participate in the program. As part of their partnership, Esperanza Coop and RAMAC each pays half the cost of all the field and extension agents/advisors. They provide training sessions related to bean production, monitoring activities, program promotion, and corresponding extension services.

Technical sheet

The Frijol Nica program aims to increase yield and improve crop management in general. A technical sheet showing when and how to apply fertilizers and pesticides, developed jointly by Syngenta and RAMAC, is provided to all the farmers. Before joining the program, farmers used to apply fertilizer several times a week, often unnecessarily. Now they apply a set amount according to the technical sheet, saving both money and time. The technical sheet specifies clearly when to use what amounts of which products. The program seeks to teach farmers how to improve their bean production by investing in agricultural inputs. The idea is not to give farmers inputs free of charge, but to demonstrate how to work on their own land. In addition, Syngenta launched a program called "El Espantapájaros" to teach youth about safe and scientific use of agrochemicals, with the intention that they transfer this knowledge to their parents.

Financial arrangements

Through the Frijol Nica program, producers receive all agricultural inputs from distributors. Some of the inputs are from Syngenta, while seeds and fertilizer products come from other companies. Bean production takes on average about 120 days. Syngenta, together with RAMAC, offers an integrated bundle of inputs to cooperatives on 120–150 days' credit. The credit is backed by cooperatives and financed by RAMAC, which buys Syngenta's products on 180-day terms. Bean growers who belong to these cooperatives can buy seeds, herbicides, and fertilizers and repay the loan when the bean harvest is sold. The cooperatives are responsible for distributing the inputs and collecting the money.

RAMAC sell inputs to Frijol Nica producers at prices 6% lower than the market. Producers can access these inputs through Esperanza Coop or other cooperatives to which they belong, either on credit, in cash, or by paying in beans. Transportation costs of both inputs and beans are shared by the producers.

The price of black beans offered to producers is determined before the beans are planted. The basic price was set at US$320 per ton, and final prices may be adjusted slightly according to the market. Costs to transport the beans are usually assumed by the producers.

The Frijol Nica program runs with several formal cooperatives, including Esperanza Coop, ECOPMAVSA (El Tigre), Sacacli Coop, and Esquipulas Coop. In 2013, some ECOPMAVSA growers received credit from Banpro, the largest bank in Nicaragua, which works with producers who are organized in cooperatives. Banpro is the first bank to provide financial support to bean producers. The bank offers relatively low interest rates (around 10% per year) and gives credit for 80–100% of the total cost of inputs specified in the technical sheet.

Value created by the Frijol Nica program

The program elevated bean producers' self-esteem and motivation and helped transform subsistence bean production into a business. Growers have seen yield increases of up to 100%, helping to improve their incomes and lift them out of poverty. Some have been able to obtain credit from banks for the first time. These benefits have boosted the program's growth, in both total area and the number of farmers it works with. It has increased from 300 farmers with 1,400 hectares in 2007 to 6,500 producers operating on 13,000 hectares in 2012. Another indicator of the impact of technology can be seen from the fact that 100% of the produce was grown using certified seeds.

Benefits realized by the farmers

Marcelino Valenzuela Altamirano, a farmer from El Tigre village in Jinotega, has been with the Frijol Nica program for over 3 years. He is a member of the ECOPMAVSA Cooperative. He has a total of 7 hectares of land, of which 2.8 hectares were used for black bean production and 0.35 hectares for red beans. For his bean production, he got a loan from the ECOPMAVSA Cooperative, which helped him to purchase inputs from RAMAC and Syngenta. Mr. Valenzuela received extension services from RAMAC's field advisors, who visited his farm regularly and offered him training and field days. He followed the instructions on the technical sheet when applying agricultural inputs. He achieved a yield of 3.6 tons per hectare for red beans and 3.6–4.3 tons per hectare for black beans in 2012. All of his black beans were sold at a price of US$333 per ton to the Esperanza Coop, which exports overseas. All the red beans were produced for home consumption. The total cost of bean production was about US$595 per hectare. As a result, Mr. Valenzuela made a net profit of over US$600 per hectare, amounting to approximately US$1,700 from black beans per crop cycle. Since he joined the program, he has seen a clear increase in yields, from 1.7 tons per hectare to nearly 4 tons per hectare. In addition, his bean production has been better organized in terms of input supply and application and market channels.

Jaime Martin Gutierrez Riso, a producer from Sacacli village in Jinotega, has been with the program for a year. As a member of the Sacacli Cooperative, which works with Frijol Nica, he received agricultural inputs and technical support from RAMAC. He has 3.85 hectares, farming 2.8 of black beans and 0.7 of red. In 2012, he sold all the black beans for export and half the red beans to the domestic market. Mr. Gutierrez receives technical assistance twice a month from RAMAC, and his cooperative and obtains post-harvest services from Caritas and Banco Semillas. He achieved a yield of 4.3–6.4 tons per hectare for black beans and 2.1–2.9 tons per hectare for red bean seed production. He sold the black beans at US$333 per ton and red beans at US$250 per ton. With a production cost of US$476 per hectare for black beans, he made a net income over US$3,000 per crop season in 2012. Mr. Gutierrez experienced yield increases for both red and black beans after joining the program.

Leonidas del Carmen García Vargas, an independent farmer from Señor de Esquipulas village in Matagalpa, has been with the program for 2 years. He owns 1.4 hectares, of which 0.7 are used for black bean seed production. He got a loan for inputs and technical assistance from RAMAC and Esperanza Coop. He achieved a yield of 3.3 tons per hectare and sold all the seeds to Esperanza Coop at US$420 per ton. With a production cost of US$595 per hectare, he earned a net income of US$542 per crop cycle.

ANALYSIS OF THE PRIVATE EXTENSION WITH RESPECT TO RELEVANCE, EFFICIENCY, EFFECTIVENESS, EQUITY, SUSTAINABILITY, AND IMPACT

The Syngenta Frijol Nica program was specifically designed to address the constraints faced by bean growers in Nicaragua, in particular a lack of credit and technical assistance. The program is therefore highly relevant for bean producers, as shown by its increasing number of participants. Before the program, producers either applied agricultural inputs insufficiently or excessively, and at the wrong times. Some farmers occasionally received extension services from NGOs, but they were not directly targeted at input supply and application for bean production. The Frijol Nica program has not only introduced advanced inputs such as certified seeds and crop protection products, but also the technical sheet that clearly specifies optimal input application at different growing stages. Technical training and farm visits further help farmers to address emerging issues in the field.

The program has proved to be efficient and effective, in the sense that it has increased farm productivity and income in a relatively short period. In addition, the program builds on the strengths of each partner to maximize the impact at the farm level. It combines the strengths of Syngenta, a technology leader in crop protection, with those of RAMAC, the local distributor of inputs, and Esperanza Coop, which distributes certified seeds and deals with export markets, providing farmers with bundled services from input supply through to output marketing. The partnerships have clearly enhanced the efficiency of the program. Moreover, associated training schemes have ensured a smooth flow of information and knowledge between Syngenta field advisors, their counterparts in RAMAC, and district cooperatives, as well as farmers. The extension approach functions well and helps to efficiently resolve any issues on the ground as they emerge.

The program is also equitable; it does not select individual producers. It works with all of the producers in the partner cooperatives who are willing to participate in the program. There are also independent growers, who can join the program on their own, but they have limited access to credit since credit is routed through the cooperatives. Once in the program, producers are all treated in the same way, receiving advice and technical support from field technicians and various seminars.

The program seems to function sustainably, and this is reflected in several areas. First, the program has brought value to all participating parties (i.e., the farmers, cooperatives, RAMAC, and Syngenta). As long as this continues to be a mutually beneficial relationship, the program is likely to continue. Second, the program's innovative financial arrangement provides opportunity for farmers to access credit for inputs from their cooperatives. This aspect continues to be highly attractive for farmers to participate in the program. Third, the cost of extension is shared by the private partners, who can still turn a profit, which indicates the financial sustainability of the model. The program's long-term sustainability depends on how the market environment for red and black beans will evolve. In general, farmers prefer to grow the crop with the better market price, which, in recent years, has not been obvious.

With regard to impact, the Syngenta extension model has proved successful in bringing benefits to the farmers involved and to their communities. Increases in productivity and income have been remarkable. The adoption of new technologies in bean production has had spillover benefits for other farming activities. Farmers increasingly view bean production as a business. Moreover, their positive experience has boosted their confidence as farmers and motivated them to continue farming. The relationships that farmers have built and the knowledge they have gained has also impacted farmers' wives and sons. They have started to see that agriculture is a good economic option for their families and their communities as a whole. In addition, the Frijol Nica program works through partnerships to help these communities to improve access to and quality of water as well as their housing conditions and biodiversity.

Another interesting activity developed under this program is the facilitation of farming opportunities for landless farmers using strategies such as "Mediador": Large-scale producers lend land to landless farmers on the condition that they share the profit and some of the input costs. For example, Rolando Payano, a producer who owns 46 hectares, helped 15 other producers to cultivate beans and generate income from his land. They saw how yields increased after receiving production guidance and technical assistance from the program. Mr. Payano received a special award from RAMAC in 2013.

COMPARISON WITH PUBLIC EXTENSION AND OTHER PRIVATE SECTORS IN THE COUNTRY

Agricultural extension in Nicaragua is oriented toward enhancing farm productivity and creating food security in rural areas. The main beneficiaries of extension services are small-scale farmers and their families who have little or no access to technologies and/or markets. Extension providers include the public-sector entities such as INTA, private companies, and a variety of NGOs and other organizations. Cash crops receive the most attention from extension officers, but awareness of the need to support local food crops is increasing.

INTA is the main public institution responsible for generation and diffusion of innovation in the agricultural sector. It has been through a series of major policy and institutional reforms since the 1990s. Over the years, INTA has applied different methods aimed at developing a demand-driven and client-focused extension system. It started with the Massive Public Technical Assistance Program (ATPm) in 1995, and later adopted the Public Technical Assistance Programs (ATP 1 and ATP 2), under which cost-sharing strategies for financing extension services were sought through various partnerships including private companies. These programs served around 41,332 families, or about 20% of all those counted in the 2001 Agricultural Census (Piccioni and Santucci, 2004). In 2001 the government launched a long-term investment in a four-phase agricultural technology program supported by the World Bank. This program aims to integrate public and private research, extension, education, and training into a cohesive, integrated agricultural knowledge system. The program will run for 16 years, and includes the creation of the Nicaragua Foundation for the Agricultural and Forestry Technological Development (FUNICA), an independent council with representatives from farmers' organizations, universities, NGOs, and other public and private organizations.

INTA has been working on strategies to improve crop productivity and living conditions of the rural poor and to empower small producers, women's groups, and small entrepreneurs through extension activities. One of INTA's most popular extension methods is "Escuelas de Campo" (Farmer Field Schools). In addition, the Technical Assistance Fund ("Fondo de Asistencia Tecnica," FAT) encourages the private sector to support farmers with required technical assistance. The Farmer to Farmer ("Campesino a Campesino," CaC) participatory extension method for maize, beans, and sorghum in the Boaco region has benefited about 200,000 families (Ortiz Dardón, 2009). INTA extensionists combine the ATP programs with various other methods to transfer advanced technology and knowledge to producers. The frequency of farm visits depends on the needs of the crop and farmers, but typically visits are carried out once or twice per month. The cost of extension services depends on the methodology used and the location of farmers. According to Ortiz Dardón (2009), it varies in the range of US$50–300 per service.

Private companies usually train their own extension agents and field promoters to reach out to farmers or cooperatives through farmers' forums, field day visits, seminars, and farmer field schools. In Nicaragua, the Agri-Inputs Distributors Association (ANIFODA) suggests that producers need privately delivered technical assistance and that the government should support private companies in doing so (Ortiz Dardón, 2009). The private sector has achieved positive results by using extension in agriculture. Companies such as SERVITEC, ADET, and SETAGRO significantly increased the production of grains, and the use of improved seeds and metallic silos helped producers to store the grain until prices rose (Ortiz Dardón, 2009).

With respect to beans, public-sector institutions like INTA and MAGFOR have been supporting the Frijol Nica program with training and seed certification, but only to a limited extent. INTA has offered some extension services, but it only works directly with cooperatives. Other institutions, such as the FAO, have supported the bean program via technical recommendations such as optimal sowing practices to increase productivity. The IICA Nicaragua (Inter-American Institute for Cooperation on Agriculture) has acted as a sponsor of the program at different points in time.

UNIQUENESS OF THE MODEL AND ITS VALUE ADDITION

The Frijol Nica program has made a unique contribution to the bean industry in Nicaragua. Designed to address the limitations faced by small bean growers, the program identified opportunities to turn bean cultivation into a highly technical, profitable activity for producers and their families. The extension model has a distinct feature: It combines input supply and technical assistance with financing as well as partnerships with commercial groups.

The startup and consolidation of the Frijol Nica program have been made possible due to various alliances, such as those Syngenta has with RAMAC, cooperatives, producer associations, national and international NGOs, financial entities, bean exporters, merchandisers, and other entities that participate in the bean value chain. These partnerships have strengthened the program and allowed it to grow. For example, in 2010 the program started alliances with international NGOs such as Save the Children, CARE, and OXFAM, to cover a total of 1,500 hectares of beans. Through these alliances, greater productivity and better product quality were achieved. All of these alliances have allowed Syngenta to transfer its technology to more than 500 technical agronomists, ultimately reaching 10,000 producers.

The program has created value for producers and all of the partners involved. Bean producers have, on average, doubled their yields and increased their income by at least 40% since the initiation of the program. They have been enabled to invest more and expand their land under bean production. Some larger landholders help other farmers by lending out their fields. For Syngenta, this program increased its sales considerably and enabled access to new customer groups among smallholders. It has also added value for RAMAC and other partners in various ways.

LESSONS FOR REPLICATION AND SCALING UP

This extension model demonstrates how the agricultural input industry approaches farmers and delivers various services in Central America. The model can be replicated and scaled up with other crops and in other geographical settings. It is worth noting, however, that success depends on yield gains achieved with the integrated package of inputs and services. Research and field experiments need to be conducted before adopting such an approach. Once it is clear that a crop can benefit from improved inputs or better cultivation practices, and hence yield good returns to investment in extension, the model can be adopted.

From the perspective of Syngenta, maintaining good collaboration with the Esperanza Coop is crucial to expanding the Frijol Nica program in the future. The cooperative understands the program well and is involved in the whole bean value chain. It can facilitate expansion and identify any emerging issues that might occur at any point in the production or marketing processes. Together with Esperanza Coop, solutions for issues identified can be developed at both the farm and cooperative levels. This will also enhance the expansion of such programs to other countries, not only to generate new sales, but to showcase the success of the program.

A major constraint to scaling up this program has to do with the financing capacity of the local distributor. To address this, Syngenta has been working with RAMAC and the Inter-American Development Bank since 2010. The bank gave RAMAC specific credit lines to increase its capacity to approach a greater number of growers.

The lack of support from public services hinders the capacity of training extension agents to transfer technologies. Since the program started, RAMAC has invited government institutions to support it more actively, but without much success. The additional support solicited was non-monetary; RAMAC was hoping for facilitation tools such as farmer registration and records as well as training modules in order to approach farmers more systematically.

The success of the program is demonstrated by the rapid growth in the number of participating bean producers over the years and the high rate of adoption of new technologies. Syngenta and its partners could use the same method to support producers in other Central American countries or it could apply the principle to other suitable crops in Nicaragua. In fact, a similar program has been created to address challenges in a broad variety of crops in Costa Rica and Guatemala.

CONCLUSIONS

This case study focused on the extension approach and the sustainability and impact of the Frijol Nica program of Syngenta. The private extension services are characterized by a combination of provision of input supply, financing, technical assistance, and linkages with traders. Effective partnerships and alliances with various entities are a key success factor. The program has generated excellent results on the ground; farmers earn much higher incomes from bean production, and their confidence in farming has increased. The program is looking for further opportunities to scale up and create value for more farmers in Nicaragua and elsewhere.

In a country where public extension has limited coverage, private extension has proved to be valuable in bridging the knowledge gaps of small-scale farmers. Innovative design and sound corporate management have enabled effective service delivery and creation of value for both producers and private actors. Such an approach can be replicated and expanded to other crops and to other countries by private companies in the agricultural input industry. Finding the right partners and a suitable extension model for each specific value chain are key to success.

The experience with Nicaragua's bean industry has revealed that there is huge potential for extension programs for neglected subsistence crops, if technological solutions can be made available to farmers in a cost-efficient manner. Improved seeds, better fertilizers and crop protection products, and innovative financing can go a long way in enhancing farm productivity and sustainability. This, coupled with proper market linkages, has the potential to transform a subsistence crop into a commercial one.

REFERENCES

Ortiz Dardón, R., 2009. Evolución de los servicios de extensión en Nicaragua. FAO.
Piccioni, N., Santucci, F.M., 2004. Nicaragua: the agricultural technology project. In: Rivera, W., Alex, G. (Eds.), Extension Reform for Rural Development. Agriculture and Rural Development Discussion Paper 12. Extension Reform for Rural Development, vol. 5. World Bank Publications, Washington, DC.
World Bank, 2013. World Development Indicators. <http://data.worldbank.org/data-catalog/world-development-indicators> (accessed in August 2012).

CHAPTER 8

Private Sector Participation in Agricultural Extension for Cocoa Farming in Nigeria: The Case of Multi-Trex Integrated Foods

Kolawole Adebayo[1], Suresh Chandra Babu[2], Rahman Sanusi[1] and Motunrayo Sofola[1]

[1]Department of Agricultural Economics and Farm Management, Federal University of Agriculture, Abeokuta, Nigeria
[2]International Food Policy Research Institute (IFPRI), Washington, DC, USA

Contents

INTRODUCTION

Agricultural extension has been provided primarily by the public sector in developing countries. In Nigeria, agricultural extension services have been dominated by Agricultural Development Programs (ADPs) funded through the World Bank loans, based in each of the 36 states and the Federal Capital Territory since the mid-1970s (Adebayo and Idowu, 2000). Agricultural extension agencies provide advice, information, and other support services to farmers to enable them to improve the productivity of their crop and animal production and thereby their farm and non-farm incomes. They are also key actors in implementing governments' rural development policies and programs However, due to low quality of service provision through ADPs, in the last 25 years, extension services have been provided by a variety of public, commercial, and voluntary agencies with varying objectives. For example, the public-sector organizations work toward national policy goals; commercial entities, on the other hand, are guided by profit considerations and try to achieve high quality standards in the raw materials reaching their processing plants and a level of production to meet plant and market capacity. The voluntary sector consists mainly of nongovernmental organizations (NGOs) which are usually guided by the welfare of the farm families. Despite the differences in their goals and approaches, all these entities seek to achieve their objectives by sharing knowledge to influence the decisions and practices of large numbers of rural farm households (Adebayo, 2004). This case study explores in some detail how a private cocoa processing company in Nigeria provides agricultural extension services to farmers, the replicability of the process in other value chains, and key lessons that transcend any particular value chain for the delivery of agricultural extension services.

DESCRIPTION OF THE COCOA VALUE CHAIN IN NIGERIA

Nigeria is the world's fourth largest producer of cocoa after Ivory Coast, Ghana, and Indonesia; producing around 250,000 tons per year. The value of cocoa exports from Nigeria was US$822.8 million in 2010. This represents about 35% of Nigeria's US$2.32 billion in earnings from non-oil exports in that year (ICCO, 2013). Most of Nigeria's cocoa is exported as beans. The major buyers of Nigerian cocoa are the Netherlands, the United States, Germany, Britain, and Brazil, with new markets opening up in Asia. Cocoa is Nigeria's largest agricultural export and the second largest source of foreign exchange after oil. Nigeria was one of the largest West African producers in the 1980s, before being overtaken by Ghana, Ivory Coast, and Indonesia due to its aging plantations, continuously diminishing yields, lack of high-yielding varieties of cocoa, inadequate and irregular supply of inputs, and inadequate technical/extension support. Nigeria currently has end-use cocoa processing facilities with a functional capacity of about 100,000 MT per year, according to industry estimates (Table 8.1).

However, the declining yield of cocoa in Nigeria (358 kg/ha as compared to over 800 kg/ha at some of the higher yielding farms in Ivory Coast) and the growing concern about quality in the global markets, portend great danger for the Nigerian cocoa industry in the long run if these issues are not adequately addressed. Of the 19 cocoa processing factories listed in Table 8.1, eight companies are currently functioning at only 35% of capacity. This is due partly to poor management and partly to low levels of supply of high quality cocoa to these plants. In response to this situation, the Nigerian cocoa industry is striving to revive the 10 nonfunctional factories.

Table 8.1 Status of all cocoa processing companies in Nigeria (2010)

S/N	Functional factories	Location	Year commenced	Installed capacity (MT)	Present capacity (MT)	Capital investment[a]	Annual export turnover (US$Mn)
1	Stanmark Cocoa Processing Co. Ltd	Ondo town, Ondo state	1991	15,000	6,000	₦4 billion	$34.29
2	Cocoa Product (Ile Oluji) Limited	Ile–Oluji, Ondo state	1984	30,000	9,000	₦6 billion	$51.44
3	Coop Cocoa Products/Olam Nigeria	Akure, Ondo state	1989	10,000	7,000	₦4 billion	$40.01
4	Multi–Trex Integrated Food Plc	Ibafo, Ogun state	2005	15,000	8,000	₦3 billion	$45.72
	Multi–Trex Integrated Food Plc	Factory 2		50,000	N/A	₦9 billion	N/A
5	FTN Cocoa Processor Plc	Iwo road, Ibadan Oyo state	2007	10,000	20,000	₦3 billion	$45.72
6	Tulip Cocoa (formerly Temple and Golders)	Ijebu-Mushin, Ogun state	2008	11,000	8,000	₦5 billion	$45.72
7	Cocoa Industries Limited	Ogba Ikeja	1967	10,000	NA	₦3 billion	$10.00
8	Agro Traders Cocoa Processing	Akure, Ondo state		15,000		₦5 billion	NA
	Subtotal	–	–	*166,000*	*58,000*	*₦42 billion*	*$272.90*

New factories under construction

| 9 | Alpha Cocoa Processing | Akure, Ondo state | 2009 | 10,000 | 2,500 | ₦3 billion | – |

Nonfunctional factories

| 10 | Cocoa Products Industries Ltd, | Ede, Osun state | 1985 | 30,000 | NA | ₦1.5 billion | Under renovation |
| 11 | Ebun Industries Ltd, | Ikeja-Lagos | | 15,000 | NA | ₦1.5 Billion | Closed |

(*Continued*)

Table 8.1 (Continued)

S/N	Functional factories	Location	Year commenced	Installed capacity (MT)	Present capacity (MT)	Capital investment[a]	Annual export turnover (US$Mn)
12	Esdee Foods Ltd,	Isolo-Lagos		5,000	NA	₦300 million	Closed
13	Urovers Comm. Proc. Co. Ltd,	Urualla, Imo state		10,000	NA	₦300 Million	Closed
14	Sulapan Cocoa Mills Ltd,	Ibadan, Oyo state		10,000	NA	Not available	Yet to take off
15	Marathon Group (Nig) Ltd,	Ile-Ife, Osun state		10,000	NA	Not available	Yet to take off
16	Celtic Exporters (Nig) Ltd,	Ilawe-Ekiti, Ekiti state		6,000	NA	Not available	Yet to take off
17	RMRDC Catalytic Cocoa Proc. Co.	Alade-Idanre, Ondo state		5,000	NA	₦300 million	Closed
18	Owena Mills Ltd,	Akure, Ondo state		10,000	NA	₦300 million	Closed
19	Oregun Mill	Oregun road Ikeja, Lagos		10,000	NA	₦300 million	Closed
	Subtotal	–	–	*111,000*	–	*₦4.5 billion*	–

[a]1 US$ = 164 Nigerian naira (2013).
Source: Secretary General, Federation of Agricultural Commodity Associations.

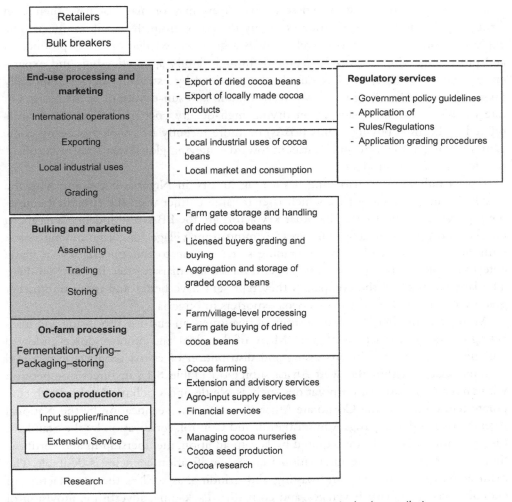

Figure 8.1 Representation of cocoa value chain in Nigeria. *Source: Author's compilation.*

The cocoa value chain in Nigeria has four stages: cocoa production, processing, bulking and marketing, and end-use processing and marketing (Figure 8.1). At the cocoa production level, support for farm input supply, extension services to support uptake of good agronomic and farm management practices, adaptive research as well as coordination and management of farmers' associations are essential. The extent to which each cocoa farmer receives and takes advantage of these services is generally a key determinant of his/her level of cocoa production and yield.

At the farm production and processing level, the major actor is the cocoa farmer, though in some locations cocoa aggregators buy cocoa from several small farms to be processed at a central location. These cocoa buyers may or may not have their own farms, but they usually have resources to buy the cocoa from the small producers. The expertise required at the farm level includes a good knowledge of cocoa production and processing, which is usually acquired through family apprenticeships and experience gained over the years, but can also be provided by extension services, especially for small farmers handling their own fermentation and sun-drying of cocoa beans. On rare occasions, extension advice on drying high-quality cocoa beans on the farm is offered by licensed buying agents (LBAs) or, more recently, by cocoa processing companies making a downstream investment toward achieving global standards or certification for their cocoa products.

Cocoa bulking and marketing has a long history in Nigeria, beginning with the establishment of cocoa boards, which later became extinct with the structural adjustment program of the 1980s. The other key actors are the LBAs and independent cocoa buyers who operate in a specific area, roaming from village to village buying cocoa without grading or with dubious grading systems. Due to this uncertainty in grade determination, the farmers on average get less than the market price for their produce. The larger LBAs and the companies they represent have better and more transparent grading systems and are the main cocoa exporters in Nigeria.

More recently, there has been a surge of investment in end-use processing and marketing of cocoa products in Nigeria. Many of the older and more resource-endowed bulking agents invested in processing and distribution of cocoa products locally and to some extent within the West Africa subregion (Table 8.1), in direct competition with larger multinational corporations such as Nestlé and Cadbury. At this level, corporate registration by the Corporate Affairs Commission, certification by the National Agency for Food and Drugs Administration and Control, certification by the Standards Organization of Nigeria are required by law. In addition, membership in and certification by at least one of the international cocoa certification agencies is desirable. The main activities at this level are roasting and grinding, as well as the manufacture of chocolate confectionary, beverages, and cosmetics. Packaging, advertising, provision of downstream support to farmers (extension services), and collaboration with research organizations (mainly the Cocoa Research Institute of Nigeria [CRIN] and universities) are the key activities that the companies engage in.

DESCRIPTION OF THE EXTENSION APPROACH

The public extension system in Nigeria is weak because the government has cut funding due to the high costs involved in maintaining the public system. Many of the extension activities implemented by state governments are funded through the ADPs, which in turn are funded by the World Bank loans. Some services are made available through local agencies independent of the state-level extension. Many extension activities, however, relate to the distribution of subsidized inputs, mainly fertilizer, as part of the Agricultural Transformation Agenda (ATA) currently being implemented in Nigeria. Smallholder farmers, for whom the extension services are intended, rarely form pressure groups to seek funding for better extension services, either at the state or federal level. Despite the weak and uncoordinated provision of extension services, government policymakers and industries expect farmers to increase production and productivity, and contribute to the growth of the agricultural economy. However, farming is increasingly becoming a knowledge-based activity. Farmers need support to access and use the available knowledge properly. As Nigeria cut its budgets for extension services, private-sector operators started providing extension services and purchasing raw materials from the farmers to whom they provide extension services. The provision of private extension services, however, is not well understood or documented. Although private funding of extension is desirable for specific commodities where there are buyback arrangements, public support will be needed to ensure extension services work for farmers growing other food and commercial crops in a sustainable and equitable way. Today, more than 70% of extension services for cocoa production in Nigeria are private because the public agricultural extension system has not been effective since the abolition of the commodity boards in the 1980s.

In Nigeria, Multi-Trex (www.multi-trexplcng.com) is one of the private-sector companies working in the cocoa value chain, providing extension to farmers. The aim of this case study is to examine an example of what the private sector is doing and what it can do better in the future with regard to the provision of extension services in Nigeria.

Multi-Trex Integrated Foods Plc, located on the Lagos–Ibadan Expressway in Warewa, Ogun state, was registered in 1990 as a cocoa bean merchant. Leveraging the 12-year cocoa commerce experience of its chief executive officer, the company at the outset bought cocoa beans from a select group of local suppliers, including a cooperative union of farmers, and processed and exported them to its customers, mainly in Europe. In 2003, the company expanded its operations to include processing of cocoa beans into semi-finished industrial products, namely cocoa butter and cocoa cake. Then it exclusively hired the facility of Nigeria's premier cocoa processing plant, Cocoa Industries Limited (CIL), Ikeja, Lagos. The partnership arrangement with CIL afforded the company the opportunity to retain its staff and prepare itself for the manufacturing business.

Multi-Trex commenced cocoa processing at its current location in October 2005. The switch from trading company to manufacturing concern demanded the restructuring of the company's operation. For example, there was a need to address the seasonal imbalance in the harvest of cocoa beans with the year-round demand for cocoa to keep the factory operating. Between 2005 and 2007, the inadequate stock of cocoa beans was one of the leading challenges faced by the company. Therefore, as part of the company's second phase of critical expansion, the key driving forces are advancing value addition and increasing access to raw materials.

Extension program content, target groups, and partners

The major emphases of the Multi-Trex extension and advisory services for cocoa farmers are presented in Table 8.2. The program is designed to target cocoa farmers and assist them to:

1. increase their productivity;
2. increase the quality of cocoa supplied through certification and good agricultural practices (GAPs);
3. increase input availability for farmers; and
4. ensure regular supply of cocoa to the company.

The extension program of the private company is designed to address the gap created by the abolition of the Cocoa Board and the resulting collapse of public extension for cocoa. Prior to 2010, when Multi-Trex began offering extension services, farmers had not put in serious efforts to produce high-quality cocoa, partly due to erosion of farmers' knowledge of cocoa production over the years. This is particularly true for the younger generation of farmers.

Further, the high price volatility in the cocoa market following the liberalization of the crop export sector in Nigeria had left farmers at the mercy of disreputable itinerant

Table 8.2 Problem areas of focus for the Multi-Trex Small Farmer Cocoa Extension Program in Nigeria

Problem	Emphasis		
	Not at all	Minor focus	Major focus
Closing technology gaps			X
Pest control			X
Natural resource management			X
Closing management gaps		X	
Providing market and input information		X	
Output marketing			X
Input supply			
Seeds			X
Fertilizers			X
Crop protection products			X
Equipment/machinery		X	
Credit (leasing, payment structures)			X
Distribute subsidized inputs to eligible beneficiaries		X	
Collecting crop and administrative data and providing associated reports	X		
Certification; quality assurance			X

Source: Author's compilation.

cocoa buyers, thus creating further disincentive to produce high-quality cocoa. Finally, farmers were not adequately organized to be able to demand fair prices. For these reasons, Multi-Trex, in collaboration with the Dutch NGO Initiatief Duurzame Handel (IDH), also known as the Sustainable Trade Initiative, started the Small Farmer Cocoa Extension Program in 2010.

The private extension by Multi-Trex depends on other public sector–supported programs in its efforts. The principal partners in such programs are the Federal Ministry of Agriculture, which helps the program to access inputs through its Growth Enhancement Support (GES) scheme, a component of the government's ATA, and CRIN, operating under a memorandum of understanding with Multi-Trex to supply seedlings to contract farmers (Table 8.3). The private extension program of Multi-Trex hires and trains local lead farmers and extension officers, who work directly with farmer cooperatives organized by Multi-Trex. Multi-Trex connects with input producers, input dealers, and other agro-industrial partners to improve farmers' access to their products in an efficient and sustainable manner. IDH, as well as other certification bodies (UTZ Certified, Fair Trade International, and Rainforest Initiative) provide the guidelines to run the extension program and the external quality controls required to ensure program integrity.

Table 8.3 Participating/collaborating/partnering sectors or agencies, and their main roles in the Multi-Trex Small Farmer Cocoa Extension Program

Partners	Main roles
Central public sector: Federal Ministry of Agriculture and Rural Development, Abuja	Helps in accessing inputs through GES; mainly makes fertilizer available to smallholder farmers
Research institutions: CRIN, Ibadan	Supplies improved seedlings to contract farmers and provides information on new varieties of cocoa
Farmer organizations: Cooperative Union of Farmers	Supports the organization of farmers into cooperatives in collaboration with Multi-Trex
Private companies: input producers; agro-industry and input dealers	Inputs are supplied through credit arrangements with local banks, which provide credit to the farmers to purchase inputs, which helps them to use high-quality inputs; input dealers are also able to connect with farmers to provide inputs on credit arrangements
NGOs: Sustainable Trade Initiative (IDH)	Support in the cocoa certification process
Others: UTZ; Fair Trade International; Rainforest Initiative	Support in the cocoa certification process

Source: Authors' compilation.

Extension program approach and design

The key feature of the Multi-Trex Small Farmer Cocoa Extension Program design is output aggregation through GAP, good environmental practice (GEP), and good social practice (GSP). These practices are international best practices designed by cocoa certification bodies and adapted in Nigeria by Multi-Trex. The program was designed by Multi-Trex in collaboration with IDH and UTZ Certified. USAID's MARKETS project provided some initial support at the beginning of the program. The frontline extension personnel in the program are the lead farmers, who are paid a monthly stipend of 3,000–5,000 naira (~US$20–35) to get them committed to full implementation of the program with direct ownership of their respective Farmer Field Schools (FFSs). The FFSs are the primary mode of delivery for extension messages, although there are a number of other approaches depending on the subject area and trainee characteristics, as seen in Table 8.4. Each participating farmer signs a contract with Multi-Trex, committing them to sell their cocoa beans to Multi-Trex; in return the farmer receives information on and training in better production, processing, and quality control.

Table 8.4 Extension approaches used to reach farmers

Type of approach	Organized by	Major focus	Scale and frequency
Farmers' training	Multi-Trex for lead farmers	Teaching best practices	Every year 3 weeks
Demonstration sites	Lead farmers	GAP	Every month
Field days (harvest)	Lead farmers	Teaching best harvest practices	Monthly visit of group extension officer to the farmers
Field visits	Multi-Trex and IDH	Checking compliance with GAP, GEP, GSP	Annual
Provision of crop guide or pamphlet	Learning materials provided in collaboration with IDH	Reminder of farm information	On need basis
Arranging exposure visits to best-practicing farmers	Lead farmers	Technology adoption—best practices	Twice a year
Free input and supervision for new farmers	Multi-Trex, IDH, and lead farmers	Free seedlings (adjusted for in the price of the cocoa)	On need basis
Joint training of farmers with other private/public actors	Multi-Trex and applicable inputs suppliers of dealers	Optimal of use of farm inputs	As necessary

Source: Author's compilation.

The main source of new technologies and approaches in the program is CRIN. This government-established research institute provides hybrid seedlings and solar dryer technology for drying the cocoa beans. CRIN also participates in training the lead farmers, who then teach the technology to groups of 20–25 farmers. The company has now reached 1,000 farmers from 47 groups (communities). Lead farmers (selected from cocoa-producing villages) are primarily responsible for forming and managing the farmers' groups with guidance from the extension officers. Each group discusses production and processing problems farmers face during FFSs organized by the lead farmers, and report them to the extension personnel.

The training schedule currently implemented by Multi-Trex is shown in Table 8.4. The company hires and trains its own program manager (one person), four subject matter specialists (SMSs), and 16 extension agents. One lead farmer per cocoa-producing community is selected by the program manager in consultation with the extension agents. The key innovation in this program is the embedding of a certification process which ensures a premium price will be paid for high-quality cocoa. Incentives for the lead farmers and extension trainers are thus structured according to the level of effort that each farmer puts in and the rate of adoption of technologies that yield high-quality cocoa.

The Multi-Trex extension program ensures that regular feedback is provided at each level of the information value chain. Lead farmers provide feedback to other farmers; extension agents provide feedback to the lead farmers and other farmers; SMSs provide feedback to extension agents and external auditors; and program managers provide feedback to external auditors and company management. This feedback structure is an integral part of the program's internal quality control. It is ensured by an internal control system (ICS) team that provides farmers with manuals and farm diaries. Farmers must record all farm activities in their diaries on a daily basis, and the diaries are checked by the company's ICS team. After this, an internal audit is done, each farm being inspected to ensure that the farmers have followed the guidelines that they have been given. Once the internal audit is complete, external auditors (from UTZ or Fair Trade International) are asked to come and complete their own certification processes. Certification is for both the farm and the processing factory. The external auditors review the internally completed documentation and then choose cocoa farmers at random for further investigation regarding compliance with GAP, GEP, and GSP. If successful, any product from the certified farms will be sealed with signs from the certification agencies.

The quality control system ensures accountability for every actor in the program. For example, farms are checked regularly by extension officers, extension workers are accountable for the adoption of technologies being promoted, and program managers are accountable for improvements in the productivity of participating farmers. Where an actor is found wanting in his or her duties, the system is set up such that failure to produce certified cocoa products also means failure to secure the premium on the ruling price of cocoa. Since this is a private-sector initiative, failure to secure the premium price reverberates through the value chain as the bonuses due to lead farmers, extension officers, SMSs, and program managers from the sale of certified cocoa are also denied.

The overriding purpose of the extension program is to revitalize cocoa farming as an industry in Nigeria and make it attractive for youths to actively participate. This is partly a social corporate responsibility action on the part of the company; but in the long run, it could bring higher profit for all the actors in the value chain and a positive corporate image for the company. The reward for lead farmers is mainly in the form of a stipend, but extension officers and program managers employed by the company are paid regular salaries.

Cost sharing and cost recovery

Program costs are covered in part by a grant from development partners (30%), in part by the premium cocoa prices, and in part by the company. In addition, each participating farmer either pays directly for the inputs he or she uses or enters into a cost recovery agreement with the company, where the cost of the inputs is deducted from the sale of the beans. The input dealers sometimes supply inputs on credit to farmers, who repay the dealers when they sell their products. The program is now also partnering with the USAID MARKETS project on an input credit scheme involving processors, exporters, chemical suppliers, and local banks. Under this scheme, USAID MARKETS, Multi-Trex, and the suppliers guarantee parts of the loan, and farmers get inputs on credit with a payment of 20% of the amount owed. The local bank recovers the loan when Multi-Trex buys the cocoa and repays the bank on behalf of the farmers.

ANALYSIS OF THE PRIVATE EXTENSION WITH RESPECT TO RELEVANCE, EFFICIENCY, EFFECTIVENESS, EQUITY, SUSTAINABILITY, AND IMPACT

Perspectives of the producers

Farmers and traders claim that government policies have been unfriendly to the cocoa industry, especially since the abolition of the Cocoa Marketing Board, and they wish that the Board would be re-established by the government. Farmers claim that the government is not sufficiently supportive of cocoa cultivation, particularly with respect to increasing access to credit and inputs including agrochemicals. Traders claim that state governments levy excessively high charges—US$5 per ton grading fee and US$18.75 per ton produce duty—in addition to taxes, such as taxes on haulage per ton, particularly in the states of Oyo, Ogun, and Lagos. The farmers and traders claim that because of the myriad problems confronting the cocoa sector, young people are not attracted to the business; most cocoa farmers are aging, and unless appropriate actions are taken the cocoa production sector in the country will be adversely affected. There is a need to link those youth in the cocoa production and processing process with the private extension system to involve young farm leaders as lead farmers as one way to attract youth to cocoa farming.

The private extension system tries to mitigate several challenges confronting small-holder cocoa producers. According to many cocoa farmers, the state governments' control of inputs is unsatisfactory and input prices are hence unstable, particularly agrochemicals and other government-subsidized inputs. Most agrochemicals are not available at critical moments in the production cycle. For example, Ridomil, a fungicide, is sold at a higher rate through the government subsidy scheme (US\$1.13–1.56) than fungicides labeled "not for sale" (US\$0.63–0.94). In order to reduce costs, some farmers still use Gammalin (a fungicide that has been banned by the World Health Organization). The traders complained that an Aquabol (a device used for testing the beans' moisture level) is costly for them to acquire, whereas if they are able to own a portable Aquabol it would go a long way in quality checks and control at the farm gate level. Thus improving the quality and quantity of cocoa production in the smallholder sector depends crucially on improved extension and input services.

At present, private extension services are tailored to meet certification requirements of cocoa production and processing. However, farm-level innovations have not been incorporated into extension services; doing so could benefit other farmers and help increase efficiency in reaching program targets. For example, in order to earn additional income, some farmers have improved indigenous technologies to add value to cocoa by-products (such as using cocoa seed in soap making) and to improve drying processes. Another area where extension services could be improved is related to the drying process. In order to increase the quality of cocoa beans processed at the farm level, farmers have adapted three types of dryers:

1. *Solar dryer.* This is a raised platform supported by four stakes, on which cocoa beans are spread for drying. The structure is covered by a weather-proof, transparent, polyethylene sheet. It is cheap and easy to construct, but dependent on sunlight.
2. *Chimney dryer.* This is typically a mud or concrete structure with a heating compartment where firewood is burned to generate heat to dry the cocoa. It can be used when sunlight intensity is low, but the smoke can easily contaminate or "cook" the beans if mishandled.
3. *Pipeline dryer.* This uses drums and pipes. Water is boiled in the drum and the steam is channeled through the pipes to generate heat to dry the cocoa beans. The heat is easy to control and the smoke does not come in contact with the beans, but construction of the system is complex and expensive for a smallholder.

All these methods of drying could be improved through farm-level innovation. Public–private partnerships could help in this context. As one farmer observed, dried cocoa pod husks burn slowly and steadily, so some farmers now use dried husks as fuel to power any variants on the dryers mentioned above. The farmers believe that government (or other stakeholder) intervention is needed to efficiently scale up these technologies and to introduce automation in the dryers.

Some farmers have complained of dwindling harvests over the years. For example, a farmer who previously produced 1.875 tons/ha is now producing fewer than 0.469 tons/ha. This could be because of the age of the cocoa plantations and the abandonment of best production practices learned in the FFSs and Farmer Business Schools. The farmers claimed that the expected support from government was not forthcoming, and hence the cooperatives formed at the end of training died a natural death. However, the farmers noted that CRIN training on organo-mineral fertilizer was found by the farmers to be useful, not only because it helped them to dispose of farm waste efficiently, but also because it increased the yields of other crops they are growing.

According to some farmers in the state of Osun, it is difficult to access seedlings of improved cocoa varieties, and when they attempted to procure the improved varieties from CRIN, they were told that the Institute cannot sell small quantities of seedlings. This is an area in which the private extension could help. Furthermore, during the off-season, agrochemicals are relatively cheap, but farmers lack the funds to purchase these inputs in advance. Coupled with the higher prices in the high demand season is the adulteration of agrochemicals, with negative consequences on the profitability of cocoa farming.

According to the local traders, they and the small farmers from whom they purchase cocoa are exploited by local buying agents and exporting firms because their size allows them to obtain more capital and hence more power. Furthermore, the local traders revealed that banks either refuse or are reluctant to extend credit to rural dwellers (including cocoa traders and farmers) because the banks place little value on rural assets (particularly land). The local traders also complained that even the little credit available usually comes with stringent conditions, including high interest rates and a very short repayment period.

The farmers who participate in the Multi-Trex Cocoa Extension Program explained that despite the achievement of the program in improving the yield and quality of cocoa, one element still threatens the ability of farmers to produce cocoa that meets international standards: the changing climate. Changing climatic conditions, such as continuous rainfall during the harvest period, prevent the farmers from properly drying their cocoa. This has adversely affected the entire cocoa value chain. Developing innovations for adapting to the changing weather patterns will help farmers meet the production and processing targets.

Perspectives of the extension workers

According to government officials in Osun State Ministry of Agriculture and Food Security, the state has continuously made improved variety seedlings available to farmers, initially free of charge but now at a rate of ₦10/seedling. In 2014, the government was set to distribute 5 million cocoa seedlings to farmers. The seedlings are sourced from seed gardens in Esa Oke and Ilesha. A third seed garden is in the process of establishment (in collaboration with CRIN) in Ile-Ife. It was observed that clustering or associations of cocoa traders were nonexistent in Oyo state, leading to lack of quality control and compliance with government policy and regulations. With limited government control and interest in the cocoa sector, cocoa associations could help farmers and traders maintain business integrity and provide a voice in issues of interest to their members. The traders opined that re-establishing the marketing board would go a long way toward regulating and enforcing compliance. The experience of the Ghana Cocoa Board has shown some positive results in helping farmers. However, the transaction costs could be high in establishing such boards if they do not perform well.

The Multi-Trex extension officers have found that more young farmers are getting involved in cocoa production under the extension program established by the company. They find that the farmers are motivated by the agribusiness focus of the program and the visibility of the link between what they do in their respective villages and the international market for cocoa and cocoa products. In addition, they opined that younger farmers are eager to take advantage of the available training. Most farmers could see a direct link between their training and their farm productivity and competitive advantage gained from adopting the lessons in the training. Business orientation, use of information and communication technology, and connecting to regional and global markets are the key attractions for youth in cocoa production and processing. Private extension systems should focus on these aspects in order to attract youth to cocoa production.

The extension officers also noted that farmers diligently watch for sunlight and always stay around to spread the beans and pack immediately if there are signs of rain. For this reason, it is important to explore alternative means of cocoa drying. This is an area in which the program does not conduct any research and innovation currently. At present, CRIN is the program's primary research partner. Resources for CRIN and other relevant institutions (such as universities) to research alternate drying methods are needed from the public sector in order for the private sector extension to be sustainable and innovation oriented.

Finally, creating rural infrastructure for better transportation and increasing the educational level of the farmers will contribute development of the sector in the long run. For example, the current extension program does not provide support in key areas such as farm transportation. Farmers resort to using motorcycles to collect and transport harvested cocoa, which requires hard labor. The extension officers suggest that state governments need to promote cocoa farms (cocoa model village settlements) to encourage young farmers. The village should come with adequate amenities, including a good road network. They also suggest that a regular farmer training and retraining program needs to be established. At present, developing educational programs for the farming community involving production commodities other than cocoa is beyond the resources of the Multi-Trex program.

Perspectives of company management

In the Nigerian cocoa sector, more than 70% of extension services are private. They are mostly provided by companies that engage the farmers to sell inputs or buy the output. International cocoa buyers, through the World Cocoa Foundation, the International Union for Conservation of Nature, and the Cocoa Producers Alliance, initiated a sustainability forum to improve cocoa production and ensure GAPs in Nigeria. This initiative attracted the private sector to make greater investments in ensuring that high-quality cocoa is produced, from the farmers through to the cocoa processing factories.

A few years ago, international certification for chocolate quality and adherence to social and environmentally friendly practices in cocoa farming became standard. Certification bodies were established to ensure that farmers abided by regulations on deforestation, fishing, upstream chemical spraying, use of child labor, and hazardous operations such as spraying and carrying heavy loads. Extension services are expected to conform to the Certification Capacity Enhancement curriculum put together by IDH and the Sustainable Agriculture Network. In the context of Multi-Trex, the program starts with training of trainers (TOT) for 40 lead farmers, who are expected to then train 20–25 farmers each. The TOT is 3-week residential training program. Participants will then return to their groups to train them in the form of FFS for 4–5 months. These are taught by the lead farmers in GAP, GEP, and GSP. All major cocoa exporters in Nigeria are following this method. Multi-Trex is the biggest exporter of cocoa, handling 65,000 tons per annum.

As a matter of courtesy, when the company works in a state of the Federation, the state ADP, a major functionary that is supposed to deliver advisory service at the state level, is informed of the company's activities. CRIN trains the staff of the company and provides technical support for seed gardens for hybrid varieties. The current varieties mature in 3–4 years, whereas hybrid seedlings mature in 18 months. Additionally, the hybrid varieties are expected to yield 1–1.5 tons/ha, two to three times the yield of the regular variety. The key performance indicator is the number of tons of cocoa sold to the company. Thus the major focus of the private extension program provided by Multi-Trex is to increase the area under hybrid cocoa varieties. In general, there is high-level leadership in Multi-Trex for developing the cocoa industry in Nigeria, including partnerships for research and extension.

Perspectives of the community at large

An interactive session was held with the chairman of the Cocoa Processors Association of Nigeria (COPAN). The chairman explained that members of COPAN may only process cocoa beans by cleaning them, or they may process the beans into intermediate products such as cocoa butter, cocoa liquor, cocoa cake, and cocoa powder for the EU market. He said that domestic cocoa grinding has decreased since 1986 due to a lack of sufficient cocoa beans. This is one of the factors that were responsible for some firms moving out of Nigeria. For instance, Cargill moved to Ghana in 2009/2010. Other reasons for the firms' departure are the low quality of Nigerian cocoa beans and high producer prices. Thus the role of extension provision from the private sector is to increase both the quantity and quality of the cocoa produced in Nigeria.

Against this background, the federal government of Nigeria introduced the Export Expansion Grant (EEG) scheme as a policy tool to provide an incentive for the stimulation of export-oriented activities that will lead to significant growth in the non-oil export sector in the country. For a COPAN member, the EEG is prorated by 15–30% according to criteria such as number of staff and processing capacity. The chairman of COPAN stated that past government policies enticed processors to become involved in the cocoa value chain. However, inconsistencies in the policies, policy somersaults, and government instability, as well as grossly inadequate infrastructure, prevented the policies from yielding the expected impact. Further, the chairman said that intermediate cocoa products were charged a duty of 4.6–8.1% upon export, whereas exported cocoa beans were not taxed and the exporter was given 10% of the total export value as a grant by the government (through the EEG). With no infrastructure back-up, processors end up spending about 28–30% of the EEG from the government on running generators, maintaining facilities, and paying employees. The balance is used to cover part of the transaction costs of cocoa exportation.

Furthermore, according to the secretary general of the Federation of Agricultural Commodity Associations (and former coordinator of the Southern Zone Cocoa Association of Nigeria), the challenges confronting processors in Nigeria include:

1. Grossly insufficient power;
2. Lack of raw materials and their continuous flow for processing;
3. Insufficient financing/credit finance;
4. Limited accessibility of EEG;
5. High costs of machinery and spare parts;
6. High export tariff to Europe—2.5%; and
7. Low demand abroad for fully processed products.

Overall, all the stakeholders agree that the Multi-Trex's cocoa advisory service is efficient and effective in meeting the needs of the company. The extension ratio of 1:63 (lead farmer to farmer) as compared to 1:1,200 in the government-run extension service, is considered appropriate for the program's target. This ratio is easily manageable to target production recommendations to meeting the needs of individual farms and farm families. The sustainability of the Multi-Trex advisory service currently hinges on its cost-recovery approach, as discussed above. So long as the cocoa processing industries and related businesses continue to run at a profit, the incentive to continue the service will remain high.

COMPARISON WITH PUBLIC EXTENSION AND OTHER PRIVATE SECTORS IN NIGERIA

The intrinsic motivation of the private sector to invest in extension services in Nigeria is to enhance the quality of the cocoa beans they purchase and ensure their profitability in processing. This is the cause of the growth in private extension services in the cocoa industry. Among other factors, the policy inefficiencies of government interventions in the cocoa sector that affect the processing industry and changes in the international cocoa trade environment have catalyzed the emergence of private extension services. Although the pace of private-sector investment in extension services has been slow, it is seen by farmers as a welcome alternative to the inefficient government-run extension programs. In addition to securing the supply of raw materials needed for their operations, the idea of corporate social responsibility is gradually gaining ground, extending beyond the immediate profit motives of the processing companies.

It is noteworthy that farmers continue to request government involvement in the funding and delivery of agricultural extension services, due mainly to the facts that it is free and that they may complement the private sector in some areas that the private sector cannot deliver on its own. However, the opportunities created by private funding and management of extension services are not fully lost. Many farmers interviewed advocate that, at the very least, the government must provide supportive and regulatory services to prevent exploitation by privately run extension services. They opine that a sizable proportion of the resources currently spent by the government on agricultural and rural development programs should be channeled into financial institutions to provide relatively cheap investment resources for young entrepreneurs in both the cocoa industry and other agriculture enterprises. Thus, it is safe to say at this point that private extension services in Nigeria currently complement the not-so-effective public extension services, but do not fully replace them. *The implication of this is that public extension services need to evolve to optimize the resources that government and the private sector invest in agricultural extension services.* The model of extension services adopted by Multi-Trex can be followed for other commercially traded commodities where companies can profit from increased farmer productivity.

The key value addition in the Multi-Trex model is that the motivation to collect dried products from several small farmers pulled together through lead farmers offers a simple model for sustainably maintaining profitable cocoa farms by smallholders. Multi-Trex is currently operating an advisory service whose cost is recovered when cocoa beans are purchased from farmers. This is considered sufficiently cost-effective to continue the extension program. Even though farmers do not pay for information received directly, the current system is highly cost-effective in that it produces a win–win situation for both the farmers and the company. This is mainly because the adoption of farm technologies and advice is directly related to farm outputs and therefore the income of participating farmers. In the public extension service, this relationship is neither obvious nor clear, partly due to the wide range of services that the public sector extension is asked to deliver. Further, the public extension service is primarily focused on the objective of food security and therefore continues to promote technologies that encourage farmers to produce more food, particularly for the local economy. This fix of the public extension service and the attendant perception of cocoa farming as an export-oriented enterprise have resulted in a situation where public extension services for cocoa farming remain moribund for the cocoa farmers, and indeed for most of the smallholder farmers growing other crops.

UNIQUENESS OF THE MODEL AND VALUE ADDITION

The Multi-Trex Small Farmer Cocoa Extension Program is a new service developed primarily to meet the company's industrial needs for a regular supply of high-quality raw materials that satisfy standards for international certification. It therefore aims to deliver an extension service that increases cocoa productivity and quality, and through this pathway raises farmers' income. There are three unique features of this program:

1. It makes profitability of cocoa production for farmers a key consideration in its operation rather than simply focusing on increasing the levels of cocoa production for its own supply. The economic returns to the farmer provide an incentive for them to continue to produce following best practices, as recommended by the private extension. It also makes cocoa production more attractive to youth who are more interested in cocoa farming as an agribusiness than as a way of life.

2. The drive to generate a high-quality product to meet the requirements of international certification agencies has the additional advantage of making cocoa farming more competitive in Nigeria. Certified cocoa attracts a premium over the current cocoa price and is more attractive to a wider range of buyers than uncertified cocoa. As such, producing high-quality cocoa is driven by both the expectation of higher economic return and that of an opportunity to explore new markets. The extension system of Multi-Trex has been able to bring the actors and players together to achieve this goal.

3. The direct connection between producers and processors and end users of cocoa beans facilitates more efficient communication and keeps the extension service demand driven and focused on issues highly relevant for the smallholder cocoa production. Though this feature increases the impact of the extension service for a specific value chain, it carries a high time cost for farmers if they produce multiple crops or livestock. Conversely, in the public extension service, a single extension agent exchanges information with farmers on different crops and livestock in a holistic manner, though the quality of service may be low and have inadequate coverage. The implication of this is that in the long run, either small farmers will become more commercially oriented and concentrate their efforts on fewer crops

and/or livestock, or the nature of private extension services will evolve in ways that meet the diverse needs of small farmers rather than those of the company that promotes it. This is particularly true for the non-cocoa producers even today.

LESSONS FOR REPLICATION AND SCALING UP

The Multi-Trex Small Farmer Cocoa Extension Program has been in operation since 2010, reaching 1,000 farmers by the end of 2013. It plans to scale up the program from its current operations in Ogun to other states in southwestern Nigeria. These plans appear feasible given that cocoa farmers have not enjoyed a functional public extension service in the other cocoa-producing states for more than three decades.

The main challenge to sustaining the program currently involves the contractual arrangement. For example, the incidence of side-selling of cocoa beans after reaching an agreement with the company remains a challenge. In the current contract, farmers who want to leave the program will have to pay for inputs already acquired through the program up to the time of the exit. The program also faces some competition from the LBAs in engaging farmers; their field-level contact sometimes entices farmers to sell their cocoa beans elsewhere. Farmers who have maintained a long-term relationship with LBAs feel uncomfortable abandoning them to join the Multi-Trex Small Farmer Cocoa Extension Program.

Despite these challenges, the Multi-Trex Small Farmer Cocoa Extension Program offers a model that can be replicated by other companies because it is simple and directly rewards farmers according to the quantity of cocoa sold to the company. In addition, it improves farmers' income from the sale of cocoa beans. The Multi-Trex extension program aims to increase the self-reliance of smallholder cocoa farmers and village-level processors, as opposed to fostering total dependency on external actors. The self-reliance goal means treating smallholders and other value chain stakeholders as development partners. Integrating issues of equity, self-reliance, and sustainability on a large scale is essential, though challenging.

The Multi-Trex program has taken these factors into account by anticipating potential tensions stemming from the rapid development of cocoa value chains, from the goal of aiding smallholders, and from addressing gender disparities and sustainability. Longer-term horizons and an adaptive problem-solving approach have been used to build capacity along the entire value chain and align the key production and processing elements of the cocoa value chains.

Furthermore, the individual and organizational capacities of smallholder cocoa farmers are strengthened as they engage in more commercially oriented cocoa farming and village-level processing. Farmers with fewer resources and female members of rural communities, who need significant support in order to benefit from the cocoa value chain development, are a special target of the Multi-Trex extension program.

Strong farmer organizations allow individual smallholder farmers and processors to benefit from the value chain through collective action. Farmers' organizational capacity building takes time and resources; among the issues requiring attention are governance, trust, internal communication, transparency, and leadership. The Multi-Trex program uses a cost-effective approach to strengthening individual capacity and uses the potential of different types of farmer organizations. The program works with partners from the public, private, and NGO sectors, utilizing their respective strengths to secure resources and provide sustainable services.

Finally, the Multi-Trex extension program draws on the experiences of its staff with different relationships and networks they have cultivated over the years. These positive strengths help in the companies' efforts to scale up their model. Developing a new, inclusive smallholder value chain requires further investments in developing and nurturing the value chain relationships and aligning key actors and elements. Building strong relationships and networks along the value chain creates trust, increases understanding of interests, and clarifies expectations. The greater the social difference between value chain actors, the greater the investment needed in relationship building.

CONCLUSIONS

Private extension systems are increasingly able to fill gaps left behind by the poorly functioning public extension systems. Cocoa production by smallholders in Nigeria and the extension provided by the processing companies are typical. This case study describes issues, constraints, challenges, and possible improvements.

Nigeria was one of the largest West African cocoa producers in the 1980s before being overtaken by Ghana, Ivory Coast, and Indonesia due to its aging plantations, continuously diminishing yields, lack of high-yielding varieties of cocoa, inadequate and irregular supply of inputs, and inadequate technical/extension support. However, with the increasing demand for high-quality cocoa and the investments in high-value commodity value chains in general by both the domestic and international investors, there has been an increased role of the private sector in providing extension services in Nigeria. In the case of the Nigerian cocoa processing industry, these new investors soon realized that the public extension system in Nigeria is weak because the government has cut funding due to high opportunity costs around funding. However, cocoa farming is a knowledge-intensive activity for which farmers need support to access and properly use the available knowledge.

As Nigeria's public extension services through the ADPs are not adequately meeting the information needs of smallholder cocoa farmers, private-sector operators are starting to provide extension services. The provision of private extension services, however, is not well understood.

Multi-Trex is one of the private-sector companies working in the cocoa value chain, providing extension to farmers. This case study has provided an opportunity to understand private sector extension from the perspective of a company that intends to improve the productivity of the cocoa producers, increase the quality of the cocoa beans to meet international certification, improve the availability of inputs in a timely manner, and connect various actors and players in the value chain to ensure a regular supply of cocoa for processing. Unfriendly government policies toward the cocoa industry have pushed both the industry and the smallholders alike to form an alliance to help each other through private extension. The private sector is able to meet the information and input needs of the farmers that are demand driven and relevant to the farmers' needs. The extension services are delivered effectively and on time through a system of lead farmers, who play the role of connecting farmers to the extension agents on a regular basis. Cost recovery through the price paid to farmers has been win–win situation, as the price farmers receive is higher than that paid by the local buying agents. This model of extension is replicable and scalable with needed modifications to other states and other production environments.

The private-sector extension studied here has also resulted in other indirect and long-term benefits. First, it has made the smallholder cocoa farmers more business oriented. It has attracted more youth and young farmers to continue cocoa farming. It has connected credit and input markets to serve the smallholder cocoa producers. The quality consciousness of smallholder farmers has increased in the study area. The private system of extension has brought the international certifying agencies to help increase the standards of production and processing at the farm level and at the company level. The complementarity of public and private sectors in provision of extension and the needed inputs are welcome, although improvements could still be made in this area.

Although the private sector is able to provide needed extension services to the farmers in the case study analyzed in this chapter, additional effort is needed to ensure that other challenges facing the farmers, such as improving drying technology and connectivity of the farmers through better transportation facilities. Further, the monitoring of contractual arrangements is also needed to reduce exploitation of the farmers by private companies as well as to reduce the violation of contracts by the farmers.

REFERENCES

Adebayo, K., 2004. Private sector participation in agricultural extension services in Nigeria. FAMAN J. 7 (2), 7–12.

Adebayo, K., Idowu, I.A., 2000. The aftermath of the withdrawal of the World Bank funding for the Ogun State Agricultural Development Programme (OGADEP) in Nigeria. J. Sustainable Agric. 17 (2/3), 79–98.

ICCO, 2013. Quarterly Bulletin of Cocoa Statistics, vol. XXXVIII, No. 1, Cocoa year 2011/12. International Cocoa Organization, London.

CHAPTER 9

Jain Irrigation Systems Limited: Creating Shared Value for Small Onion Growers

Yuan Zhou

Syngenta Foundation for Sustainable Agriculture, Basel, Switzerland

Contents

INTRODUCTION

India is the world's second most populous country, home to over 1.2 billion people, of whom nearly 70% live in rural areas. Agriculture accounted for 17.5% of its 2011 GDP and employed about half of the country's total workforce (World Bank, 2013).

Knowledge Driven Development.
DOI: http://dx.doi.org/10.1016/B978-0-12-802231-3.00009-7

Key crops include rice, wheat, tea, cotton, sugarcane, and a large variety of fruits, vegetables, grains, and pulses. Globally, India was the largest producer of livestock, milk, and many fresh fruits and vegetables, and the second largest producer of wheat and rice (FAO, 2011).

Most farmers are small-scale growers, with 86% operating on farms of less than 2 ha. The average size of land holdings declined considerably between 1970 and 2003, from 2.3 to 1.06 ha (NSSO, 2006). Small and marginal farmers now work on about 42% of the total farmland operated. Of the 140 million hectares of land cultivated in 2011, 78 million hectares were dependent on rain to grow crops; the remaining 62 million hectares were irrigated (Jain Irrigation Systems Limited, 2011). Most Indian farmers remain heavily dependent on rain to irrigate their crops, most of which falls during the monsoon season, from June to September. Fluctuations in the monsoon damage crops through drought, flooding, pests, and diseases. This is reflected in the year-to-year fluctuations in the production and productivity of crops.

For many smallholders, rice provides some food security, but in order to achieve more secure income levels they need to grow higher-value crops (e.g., vegetables and fruits). Small farmers are able to grow vegetables efficiently; farmers can fetch good returns from very small farm sizes, provided the farmer has some access to irrigation. However, as the cost of inputs increases, growing vegetables is more risky than growing rice. In addition, cultivation of vegetables requires more advanced knowledge about quality inputs and their optimal application, post-harvest management, risk reduction, prices and markets, and sometimes safety and quality standards. Access to general extension is limited for small farmers, however, and desirable cash crops require far more intensive advisory services.

Jain Irrigation System Limited's (JISL) onion program exemplifies how the private sector is stepping up its efforts to integrate small farmers into value chains. JISL was the pioneer of the Indian micro-irrigation industry, later diversifying into various agribusinesses and renewable energy. In 1995, JISL started to develop improved varieties of white onion suitable for local conditions and for dehydration processing. It began a contract farming scheme in 2001 and now works with 5,000 onion farmers in Maharashtra and Madhya Pradesh. In the onion program, JISL plays multiple roles. In addition to supplying seeds, micro-irrigation, and mechanization, the company also serves as an extension provider and output buyer. The private extension provided by JISL has increased farmers' return on investment significantly and enhanced their farming capabilities.

This paper focuses on the onion value chain in India and describes the extension JISL provides to smallholders. The relevance, efficiency, sustainability, and impact of the private extension service will be discussed, followed by lessons for replication and scaling up, and concluding remarks.

INDIA'S ONION VALUE CHAIN

Key trends in production and export

Onions and other allium species, such as garlic, form one of the most important vegetable crop value chains in India. The country is the world's second largest producer after China. In 2011–2012, 15.75 million tons of onion were grown on 1.04 million hectares (Table 9.1). The area under onion cultivation has quadrupled since 1980, with the largest increase since 2010. Yield has also improved considerably over the last 30 years, but Indian productivity still lags significantly behind other countries. The 2011–2012 figure was roughly 30% of that in Israel, which has the world's highest yield, according to the FAO.

Onion is cultivated over almost all of India. The major producing states are Maharashtra, Karnataka, Madhya Pradesh, Gujarat, Rajasthan, Andhra Pradesh, Uttar Pradesh, Orissa, and Tamil Nadu. Maharashtra is the leading producer, with a contribution of 32.6% of total onion production, followed by Karnataka (17.6%), Gujarat (10%), and Bihar (7%).

India has been an exporter of fresh onions since independence. There has been an increase in the volume and value of exported onion, but exports are subject to wide fluctuations from year to year. This can be attributed to the fact that onion exports were not free, but rather were regulated by the National Agricultural Cooperative Marketing Federation, and more recently other agencies (Chengappa et al., 2012). India exported 1.55 million tons in 2011–2012, nearly 10% of total production.

Table 9.1 Area, production, and productivity of onions in India (1980–1981 to 2011–2012)

Year	Area (million ha)	Production (million tons)	Yield (kg/ha)
1980–1981	0.25	2.5	9,961
1990–1991	0.30	3.23	10,686
2000–2001	0.42	4.55	10,786
2005–2006	0.66	8.68	13,118
2006–2007	0.70	8.89	12,655
2007–2008	0.70	9.14	12,974
2008–2009	0.83	13.59	16,260
2009–2010	0.76	12.19	16,039
2010–2011	1.06	15.12	14,264
2011–2012[a]	1.04	15.75	15,106
CGR (%) (2000–2001 to 2011–2012)	11.5	16.5	4.5

Source: From Chengappa et al. (2012).

Prices for onions are more volatile than for many other agricultural products due to low elasticity of price and income and inherently unstable production. The changes in prices have a significant impact on food security, and farmer and consumer welfare. An increase in the price of onions affects household food consumption budgets, while any decrease in onion prices below the cost of cultivation affects the producer.

Varieties of Indian onions

The bulk of Indian production is red onion for fresh market sale. Local red onions are not suitable for dehydration, primarily due to poor quality, low productivity, low solids, low pungency level, and high reducing sugars (JISL, 2008). Traditional varieties of white onions are grown only in a few districts of Maharashtra and Gujarat, but they possess low total soluble solids (<13%). These varieties can be used for dehydration but these have some disadvantages. Their low soluble solids content (<13%) means that powder recovery from dehydration is limited. A 40% bolting rate during bulb production results in poor quality of the dehydrated product.

Dehydrated onions are mostly white, having high solids, often above 17% total soluble solids, or even up to 22–26% in some hybrids. They have comparatively low moisture content (<84%), and are round in shape, having a small root base with a minimum 70 mm diameter (JISL, 2008). These onions usually have a longer shelf life, free from diseases. Most white varieties and hybrids suitable for dehydration are cultivated in Europe and North America. JISL started its own breeding program in 1995 to develop varieties suitable for short-day tropical conditions in India. The Jain V-12 variety was selected as one of the best performing varieties meeting dehydration requirements.

The value chain partners of the onion sector

In the domestic fresh market, several parties are involved in the red onion value chain. There are farmers who grow the crop, traders and wholesalers who source from farmers and trade in the marketplace, consumers, and government agencies that regulate the market. There are four broad marketing channels: (1) direct to consumers, (2) through private wholesalers and retailers, (3) through public agencies (regulated markets) or cooperatives, and (4) through processors.

Smallholder farmers

Most onion growers have very small land holdings. They have little say in their onions' final market price. Farmers generally refer to the local market's prices for onion.

Traders

For historical and financial reasons, large storage capacities for onion have been owned by private traders. Traders buy storage lots and provide sorted and graded produce to retailers or buyers. Unlike farmers, traders usually compare onion prices in different markets, including export markets, before deciding where to send different grades of produce to maximize profits.

Governments (market regulations)

The Maharashtra Agricultural Produce Marketing Act (1963) was amended in 2007 to promote competitive marketing. Rules were altered to give greater freedom to farmers to sell their produce directly to consumers, processors, or manufacturers. Provisions were made to designate certain markets as special commodity markets according to the volume of arrivals, turnover, and geographical area. This is to encourage the development of specialized markets with modern infrastructure and storage facilities and private sector participation. This is a great step toward promoting efficiency in the onion markets. In Karnataka, the Act has clear provisions for increasing equity in the sale of agricultural produce, providing marketing facilities and dispute settlement, using market funds for in-house facilities, and making credit available to farmers.

The value chain for dehydrated onion products is different. Under contracts with JISL, farmers grow one of JISL's white onion varieties and then sell all the produce that meets quality specifications back to the company at either a guaranteed price or the market price, whichever is higher on the day of the sale. JISL processes the onions in its own dehydration plants and sells the product to overseas consumers. The value chain does not include intermediary traders or middlemen, although rejected produce will ultimately enter the local fresh market.

EXTENSION METHODS OF JISL'S ONION PROGRAM

JISL is a multinational organization headquartered in Jalgaon, Maharashtra, India. It derives its name from the pioneering work it did in the micro-irrigation industry in India. However, since its earlier irrigation work, the company has extended its activities into hi-tech agro-related ventures such as tissue culture plants, agriculture and fruit processing, bio-fertilizer, and bio-energy. Additionally, its operations now cover piping products, plastic sheets, and solar energy (Figure 9.1). JISL employs over 7,500 workers worldwide, including more than 6,000 in India. It is the largest private-sector employer of agricultural professionals in the country.

JISL has also diversified into food processing and has a number of state-of-the-art facilities for processing vegetables and fruits, in addition to its own research and development farm in India. It is the only private-sector company recognized by the government for agricultural activities and experiments on various agronomic and irrigation practices in line with international practices.

JISL is currently the largest processor of onions in India and the third largest in the world. The company has world-class, large-scale integrated dehydration facilities. It commissioned its first onion dehydration plant over 10 years ago, geared entirely to export.

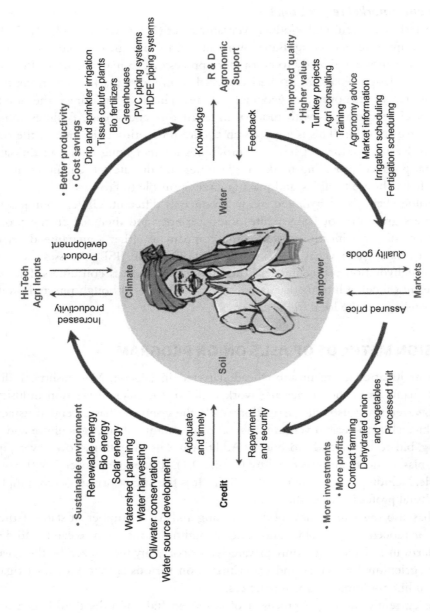

Figure 9.1 JISL's sustainable agricultural cycle. *Source: From JISL (2013).*

Today, the facilities process 120,000 metric tons of raw onions yearly, yielding 15,000–18,000 tons of finished products. The range of processed products varies in line with customer needs. The processing plant follows ISO 2000, GMP, and HACCP standards, as well as other quality management guidelines.

The company entered into contract farming in 2001. Its main objectives are to ensure a consistent supply of onion bulbs, enhance productivity and uniformity with better agricultural practices, and improve the quality of onion bulbs. The raw materials undergo quality inspection, sorting, and dehydration, ensuring maximum retention of flavor, aroma, color, and taste. Dehydrated flakes are then carefully milled into different fractions at the company's own modern facility, ensuring compliance with strict food safety and hygiene standards. During milling, the products undergo metal-detection and color-sorting at the last stage, and are then packed. These are then stored at low temperature and humidity, and supplied to customers all over the world.

Extension model

In 1995, JISL started its own breeding program to develop white onion varieties suitable for local agroecological conditions and dehydration. The breeding programs and trials ran in Bhutan and India. Jain V-12 is the variety developed and used for commercial seed production and demonstration in Maharashtra. JISL currently produces 30 tons of seeds per year and maintains a buffer stock.

JISL's onion contract farming program started in 2001 with about 100 villages and 435 farmers. At present, there are about 5,000 farmers under contract, covering 7,000 acres of land and producing 70,000–80,000 metric tons of onions. Most of those producers under contract are smallholder farmers. The average yield is about 13–14 tons per acre (far above the average 5 tons per acre for traditional varieties), and under optimal conditions can reach 26 tons per acre.

Criteria for selection of farmers

Onion farms within 200 km of Jalgaon are preferred for easy accessibility. The contract farmers are currently located in the districts of Jalgaon, Dhule, Nandurbar, and Buldhana in Maharashtra, and in the districts of Badwani and Khandwa in Madhya Pradesh. Other criteria for participation in the program include:

- Farmers have their own land, and it has fertile soil suitable for onion cultivation
- Availability of irrigation
- Availability of family labor
- Farmers' willingness to adopt modern hi-tech precision farming practices
- Experience in onion cultivation, competence and reliability of the farmers, and their ability to cooperate with others

Contract farming management model

JISL has a team of 60 well-trained young agronomists, called Jain Gram Sevaks, in its contract farming scheme. They are involved in the provision of extension services, input supplies including premium seeds, and advisory services for the developing nursery, transplanting, fertilizer and pesticide application, harvesting, proper packaging, and transportation. Before the sowing season, the agronomists organize agricultural fairs ("mela") as a platform for farmers to exchange views and experience. New technologies in onion production are demonstrated in farmers' fields as well as on JISL's own R&D farm.

On average, each Jain Gram Sevak is in charge of 40–50 farmers and about 100 acres of onions. This person either comes from the village assigned to him/her or stays there. The Jain Gram Sevaks are provided with housing, a motorcycle, a mobile phone, and a digital camera. They are the key link between the company and the farmers, and usually possess first-hand knowledge of the onion crop. They are responsible for identifying potential contract farmers, providing seeds at the right time, and advising and monitoring every stage of cultivation.

The Jain Gram Sevaks work at the village level and report to agricultural supervisors responsible for a block of villages. One supervisor monitors and guides about 7–10 Jain Gram Sevaks. Agricultural supervisors report to managers (or agricultural officers) who typically supervise four to five blocks. The management structure is illustrated in Figure 9.2. In addition, there are agricultural friends (Krishi Mitra), who are village contact persons (or messenger/lead farmer) for company operations, not only in extension, but also other areas. They usually belong to a farming family, and help coordinate farmers' access to inputs or irrigation equipment.

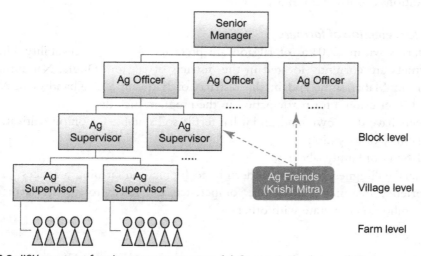

Figure 9.2 JISL's contract farming management model. *Source: Author's compilation.*

Under the contract farming program, JISL provides agricultural inputs like seeds, bio-fertilizer, micro-irrigation systems (e.g., drip), seed planters, packaging materials ("gani" bags), and a comprehensive package of extension services. Farmers can buy seeds with cash or on credit. For drip irrigation systems, farmers receive a 50% subsidy from the government and can get the other half on credit organized by JISL through existing loan schemes involving local banks (e.g., the Union Bank). JISL partners with local banks and other financial corporations to secure loans for their contract farmers. Eighty percent of JISL's contract farmers now use drip irrigation.

The company's R&D, extension wing, and contract farming team provide a complete package of practices for improved productivity and quality. The package includes information on seed planters, drip and sprinkler irrigation, fertigation and fertilizer doses, weed management, pest and disease control, and harvesting practices. After harvest, the produce is bought back by JISL at minimum support prices or market price, whichever is higher.

Onion seed planter

In recent years, JISL has developed an onion seed planter to enable direct seeding and avoid the need for heavy labor during transplanting. Using the seed planter, planting can be finished within 2 h per acre. In contrast, in traditional planting farmers need first to prepare the nursery, which takes about 6 weeks, and then transplant seedlings. More than a month is saved for cultivation by using the seed planter. The machine costs about Rs. 18,500. Because JISL entered a public–private partnership with the government of Maharashtra, farmers can receive about half this sum in subsidies. Seed planters can also be purchased jointly by a group of farmers. Renting a seed planter costs about Rs. 200–300 per acre, while manual transplanting of the same area costs Rs. 5,000–6,000. In addition, about Rs. 2,000 are needed to raise a nursery for 1 acre of sapling transplant. Rapidly rising labor costs have prompted the demand for labor-saving machines.

JAINGAP and good agricultural practices

In 2008, JISL and the International Finance Corporation (IFC) jointly carried out a diagnostic study of onion and mango supply chains to identify issues related to food safety, traceability and implementation of good agricultural practices (GAPs). The results revealed that JISL suppliers were practicing commonly noted GAPs at the farm level, but it was necessary to develop an intermediate standard to improve implementation of GAPs. As a result, JAINGAP was initiated in collaboration with the IFC in 2009.

JAINGAP is a GAP certification based on and recognized by GlobalGAP. Its objective is to ensure farmers use prescribed GAPs for sustainable production. GlobalGAP has positioned JAINGAP as an entry-level standard to effectively implement food safety management systems among small and marginal farmers. The IFC has appointed Catalyst Management Services in Bangalore to conduct monitoring and evaluation for JAINGAP.

JISL promotes sustainable agricultural practices with JAINGAP, including ground-water recharge, drip irrigation/fertigation, and resource conservation. A total of about 18,000 acres has been brought under the JAINGAP management systems, with about 7,000 onion growers complying with JAINGAP standards between 2009–2010 and 2013–2014. JISL also attempts to adhere to the Sustainable Agriculture Code (SAC), which is more comprehensive than GlobalGAP. Over 1,000 onion farmers adopted JAINGAP and SAC in 2011. JISL provides a booklet about white onion cultivation in local languages to promote sustainable production practices. It also makes available a "JAINGAP kit" of crop protection products, for which the farmers pay half the cost. The long-term aim is to cover a larger number of white onion and mango farmers under both JAINGAP and SAC. Buyers in Europe, the United States, and South America have visited some of the contract farmers to see for themselves the good practices adopted by farmers.

Purchase policy and payment arrangements

Farmers achieve increased productivity and higher income by following the recommended package of practices. JISL buys the produce, ensuring a minimum price of Rs. 4–5/kg on delivery at the processing plant. However, on the date of delivery, if the market price is higher, the company pays the prevailing market rate. If the market price is lower, contract growers are paid the minimum guaranteed price. This "double price formula" has worked extremely well. Enhanced yields with no price risk increase contract farmers' net income. This model insulates the farmers from fluctuating market prices and eliminates middlemen and associated brokerage charges.

One of the major advantages of contract farming for the company is close monitoring of onion bulb quality. Through this system, production and quality are more reliable than with open market purchases, and JISL faces less risk of losses to pests and disease. As farmers are advised well, the bulbs supplied are of superior quality and ideal for dehydration. Nearly 100% of the produce is accepted by JISL. Bulbs that are under the required size are rare, because farmers carefully select the seedlings at planting.

All contract farmers are registered and each has an account at Jain's own bank. Farmers receive checks when they sell, on which they can draw cash 10–20 days after delivery. The company accepts produce only from registered farmers. If farmers take seeds or other inputs on credit, the costs are deducted from sale proceeds at harvest. During the growing season, the Jain Gram Sevaks record farmers' production practices, including input use. Their reports go into the company's central data management system. JISL has its own software to store all the data.

Linkages with banks

Another critical factor for success is that JISL works with the nationalized Union Bank of India and State Bank of India. These banks offer crop loans for onion production and term loans for the purchase of capital equipment such as drip or sprinkler systems. JISL acts as a facilitator, liaising between the farmers and the banks. It pays back the loans to the banks directly, deducting the dues from farmers' sales to the company.

The farmers are encouraged to install a micro-irrigation/fertigation system, which requires a term loan of up to Rs. 25,000 per acre. A memorandum of understanding between JISL and the two banks sets the payback period at around 4–5 years (Jain, 2008). Contract farmers can also get crop loans to meet the cost of cultivation in the amounts of Rs. 8,500 per acre from Union Bank or 10,000 from State Bank.

Shared value created by JISL

With higher production volumes and quality yields, farmers can benefit from contract farming and realize a good return on investment. They profit from elite seeds supplied at a subsidized rate by JISL and use the latest technology and cultivation methods. Farmers also acquire knowledge of record-keeping, efficient use of farm resources, improved irrigation methods through drip/sprinkler systems, fertigation, and understanding the characteristics and demands of the processing industry. Each of them receives a free booklet in local languages on hi-tech precision onion farming. JISL has managed to convert over 80% of onion contract farmers to micro-irrigation systems. On average, farmers have achieved yield increases of 130%.

Benefits realized by the farmers

Sharad Rajarani Patil, a farmer from Shirsoli village in Jalgaon, Maharashtra, has been a JISL contractor for over 11 years. He is a typical small-scale farmer, with 4 acres of arable land. Two acres are devoted to onion contract farming with a drip and sprinkler system. Mr. Patil also grows cotton, corn, sorghum, and wheat. Onions serve as his major cash crop. He receives seeds, the micro-irrigation system, and extension services from JISL, and a Jain Gram Sevak regularly visits his field. He achieved a yield of 15–17 tons per acre during the *rabi* season in 2012 and 7–8 tons during the *kharif* season in 2011. The higher productivity during the Rabi season is attributed to better management of irrigation. The total cost of onion production is about Rs. 20,000–22,000 per acre during *rabi* and about 22,000–25,000 during *kharif*. JISL paid the market price in both seasons, namely Rs. 8.75/kg in *rabi* and Rs. 11.5/kg in *kharif*, well above the initially contracted price. As a result, Mr. Patil made a profit of over Rs. 100,000 during both seasons. He clearly sees the value created by working with JISL. On average he has earned at least Rs. 25,000 more per acre since joining the program. Onion yields have been quite stable, and JISL has absorbed most of the price fluctuation.

Our discussion with 11 farmers from Shirsoli village revealed that all of them had benefited from contract farming with JISL. The major issue that farmers usually face with onion production is the cost of labor. Manual transplanting costs at least Rs. 3,000 per acre, and harvesting about Rs. 500 per ton. The introduction of the seed planters has already reduced labor costs considerably, though some farmers say that certain aspects of the machines need improvement. To reduce costs even further, farmers would also like a harvester machine. JISL has recently developed a model which is presently under trial.

ANALYSIS OF THE PRIVATE EXTENSION WITH RESPECT TO RELEVANCE, EFFICIENCY, EFFECTIVENESS, EQUITY, SUSTAINABILITY, AND IMPACT

Contract farming is quickly evolving as an alternative marketing mechanism in India. It has the potential to combine small farm efficiencies and corporate management skills, provide assured markets, and decrease transaction costs through vertical integration. Experience in India shows that contract farming considerably increases input use efficiency through the introduction of improved technologies and better extension services (Jain, 2008). It has also allowed some farmers to access credit to finance their inputs. However, there are also many examples of unsuccessful contract farming, often related to issues such as contract enforcement, monopsony, manipulative provisions of the contracts, and delayed payments.

In Jalgaon, onion farmers have benefited from the JISL contract in a number of ways. First, the introduction of white onion varieties has been a game-changer for growers, increasing productivity and income significantly. Particularly when irrigation is easily available, white onions' good marketability makes them an attractive crop for small farmers. Second, with an assured market, farmers are insulated from selling risks and face minimal production risks because of intensive extension services. In addition, onions serve as the best rotation crop for bananas and cotton, which occupy over 60% of the cultivated land in Jalgaon. Bananas and cotton are two major *kharif* crops, while onions are mainly planted in *rabi*. In the past, farmers used to rotate bananas with wheat. Now they prefer to rotate with white onions because the return on investment is considerably higher than that of wheat.

As far as efficiency is concerned, from JISL's perspective, they have refined and improved the contract farming model to more efficiently serve the farmers and its own business operations. From the perspective of producers, the private extension is effective in the sense that they receive inputs and advice specially tailored to their onion crop. As the company buys the produce, the system operates in a closed loop. Inputs and resources are optimally used and transaction costs reduced.

The system is also very equitable. As mentioned earlier, JISL has its criteria for selecting farmers, but these are not related to gender or farm size. (However, JISL operates within the traditions of local society, whereby men own the land and make agriculture-related decisions. The program primarily contracts with males, although this is not deliberate.) The company treats farmers as friends or "part of the JISL family." The contract terms are the same for all, and the prices farmers receive depend heavily on the market rate when selling, though it does not fall below the minimum guaranteed price declared by the company before the sowing season.

The contract farming system continues to function sustainably as long as there is a market for white onions and both farmers and JISL are able to profit. Since there are no formal contracts signed by both parties, mutual trust is essential. On the financial side, the model covers the cost of extension and creates shared value for the company and farmers. Compared to many similar systems, two advantages make JISL's model stand out. First, the white onion variety provided by JISL is specifically designed for processing. As there are no other competitive processors in the area, there is little scope for poaching or side-selling. JISL's double price formula further encourages farmers not to sell to anyone else.

With respect to impact, JISL's extension model has created shared value and brought benefits directly and indirectly to the farmers and farming communities. Farmers have gained from enhanced productivity and increased income, as well as exposure to the latest cultivation and irrigation technologies. Part of the profit JISL generates through its business activities is invested in promoting the cultural, educational, and social life of the community. For example, JISL runs the Jain Gurukul training center, used by more than 40,000 farmers, students, and the government annually. In addition, JISL owns the Anubhuti chain of institutions focusing on rural and agriculture-oriented education. The Anubhuti day school in the center of Jalgaon provides free education, food, and uniforms to children from the poorest families. Two rural schools and an agricultural college have been established in the nearby village of Wakod.

COMPARISON WITH PUBLIC EXTENSION AND OTHER PRIVATE SECTORS IN THE COUNTRY

In India, agricultural extension has a mixed record of achievement. The literature recognizes agricultural extension as an important promoter of productivity increases, sustainable resource use, and agricultural development in general. But the public provision of extension has fallen short of expectations. Links between research, extension, and farmers are often seen to be absent or weak, while at the same time there are duplications of effort, with a multiplicity of agents attending to extension work without proper coordination.

Public extension is implemented at the state level in India, largely through state departments of agriculture. Many states, however, find it difficult to empower farmers with the required knowledge, technology, and input support. The extension system lacks manpower and is burdened with many tasks in addition to advisory services. This is reflected in the National Sample Survey Organization 2005 report: Only 6.4% of Indian farmers accessed information on modern agricultural technology through public extension workers. This figure rose to 40% when all sources of information were considered, including other progressive farmers.

The Agricultural Technology Management Agency (ATMA) is viewed by many as a novel approach to public extension reform. It seeks to integrate extension programs across line departments and decentralize decision making through "bottom-up" procedures that link research, extension, and farmers, as well as NGOs and the private sector. After a successful pilot program in several districts, ATMA was scaled up across the country. However, as with many other extension programs, implementation bottlenecks emerged at scale-up. These include lack of qualified manpower, technical and financial support, or a clear framework for implementing public–private partnerships (PPP). ATMA also has only weak links to other extension units such as Krishi Vigyan Kendras. The agency is currently under close scrutiny and is being reformed by the central government.

Discussions with district agricultural officers in Jalgaon revealed that most extension activities are linked with government schemes. In Jalgaon alone, there are currently 135 such schemes, which keep the public agricultural workers busy. Major extension activities include demonstrations, training, exposure visits, farmer field schools, and provision of kits (seeds, fertilizer, and pesticide) on a cost-sharing basis. In Jalgaon, there is one chief district agricultural officer, with three subdivision agricultural officers and about 15 block-level officers. In each block, there are circle-level officers who oversee two to three supervisors, each of whom has a number of agricultural field assistants. Field assistants are the lowest level public workers, having direct contact with farmers. Each assistant is responsible for one to three villages, and are mostly occupied with tasks related to different agricultural schemes. According to the agricultural officer being interviewed, the ideal ratio of public extension staff to farmers should be 1:1,000.

A majority of funding for these schemes comes from the central government, the rest from state sources. The exception is the drip irrigation subsidy, the cost of which the center and the state share equally. ATMA has one project director, two deputy directors, and two block technology managers at each block in the Jalgaon district. ATMA's main focus is technology transfer. Although a separate body, its functions are highly dependent on agricultural officers from the agricultural line departments.

In recent years, JISL has signed several PPP agreements with public institutions including the state government of Maharashtra and ICRISAT. The most recent PPP with the Maharashtra government concerns onion farmers. The government provides a 50% subsidy for onion seed planters and drip irrigation, as well as one bag of potash fertilizer per acre. Around 275 farmers had benefited from onion seed planters by May 2014. An earlier partnership between ICRISAT and JISL focused on empowering poor dryland communities through sustainable water management. In 2009, JISL signed a memorandum of understanding with the National Bank for Agriculture and Rural Development on a PPP initiative for integrated rural development. In 2011, the company signed another memorandum with the Maharashtra government to increase cotton production by small dryland farmers. This PPP will bring drip irrigation technology to 100,000 ha by 2016, which will help to triple yields.

Agricultural extension in India has become increasingly pluralistic: A large number of private, "third sector" (i.e., NGOs, foundations), and informal service providers now coexist with the public system. Market-based extension offered by agro-dealers, input suppliers, and buyers of produce is also on the rise. Prominent "one-stop shop" solutions include Mahindra Krishi Vihar and Tata Kisan Kendra. Hariyali Kisaan Bazaar seeks to provide "end-to-end agri-solutions" for farmers. Contract farming is a promising way of linking more farmers with extension and markets. Examples include schemes run by Pepsico, Field Fresh Foods, Hindustan Unilevel, and Adani Agrifresh. JISL serves as an excellent example of providing extension through contract farming, and is an innovative model of PPP.

UNIQUENESS OF THE MODEL AND ITS VALUE ADDITION

Private extension services accompanying contract farming do not always work well. Several factors contribute to the success of JISL's model. First, the company has its own R&D farm for crops like onions, pulses, oil seeds, cereals, fruits, and vegetables. Its strong R&D team has contributed to the development of improved crop varieties such as the Jain V-12 white onion, as well as the invention of the labor-saving seed planter and harvester. JISL also developed drip and sprinkler irrigation systems for small-scale farmers through its R&D program. Second, JISL has designed a suitable contract farming management model that includes agricultural workers at different levels with different responsibilities. JAINGAP was also innovative in its promotion of sustainable agricultural practices. Third, the purchase guarantee combined with the double price formula makes the scheme attractive to growers. Finally, good links with banks to facilitate access to credit also help make the model successful.

"Creating shared value" is not just a JISL slogan; it forms part of its corporate culture. Contract farming increases farmers' return on investment while creating profit for the company. The company employs hundreds of young women in its tissue culture and agro-processing plants. This employment opportunity has improved their position in a backward area.

LESSONS FOR REPLICATION AND SCALING UP

JISL's extension model can certainly be replicated in other crop value chains and geographies where appropriate, but there will be some challenges. The combined package of seeds, extension, irrigation systems, purchase guarantee, and agro-processing makes it unique and somewhat difficult for smaller companies to emulate its success. Furthermore, the current approach and the associated value chain serve a niche export market for dehydrated onion products. It creates high value addition from in-house agro-processing, which enables JISL to cover the cost of extension. For products bound for the domestic fresh market, it may be difficult to recover expenses related to private extension.

JISL recognizes the need to scale up its onion program. In the coming years, it aims to bring more acreage of the existing 5,000 contract farmers under onion cultivation. Priority will go to farmers with drip irrigation systems and within 200 km of the Jalgaon onion dehydration plant. Scaling up in other areas is not currently planned because of high transportation costs. Future expansion will also depend on whether overseas demand for dehydrated products increases.

CONCLUSIONS

This case study focused on the approach and impact of JISL's private extension program for onion production. The extension model is characterized by farming contracts, provision of improved seeds, micro-irrigation systems, seed planters, and advice on sustainable agricultural practices under JAINGAP, as well as the purchasing policy and strong financial facilitation. The innovations of the model include varietal development, invention of the seed planter and harvester, and continuous improvement of drip systems and JAINGAP. In-house R&D has served as the main source of new technologies and approaches, complemented by information gathered from seminars, conferences, and studies on relevant topics. The onion program has generated an enormous impact on the ground; farmers have increased their return on investment significantly and benefited from social welfare provisions associated with the program, such as schooling. Private extension has proven to be useful and effective in this case, with important implications for replication in India and elsewhere.

The successful vertical integration of JISL's extension program in the white onion value chain, from input provision to processing, is unusual. Only a few private companies are able to participate in the whole chain; most cover only a limited section. The JISL approach is targeted at a special value chain and niche market, and thus has limitations for emulation and replication. However, if similar niche markets are discovered, similar approaches can bring shared value to both farmers and companies.

There are complementarities between private and public approaches to extension. Private extension programs exhibit efficiency, effectiveness, and impactful results that target farmers and value chains. Public extension can and should focus on crops beyond these value chains and should cover more remote and less endowed areas and farmers. This requires better understanding of the role and responsibilities of the various extension providers—public, private, or "third sector." How best to coordinate their efforts and facilitate collaboration, and how to create effective PPPs to reach more farmers are questions that remain to be addressed.

REFERENCES

Chengappa, P.G., Manjunatha, A.V., Dimble, V., Shah, K., 2012. Competitive Assessment of Onion Markets in India. Agricultural Development and Rural Transformation Centre, Institute for Social and Economic Change, Nagarabhavi, Bangalore, India.
FAO, 2011. FAOSTAT. <www.faostat.com> (accessed in July 2013).

Jain, R.C.A., 2008. Regulation and dispute settlement in contract farming in India. In: Contract Farming in India: A Resource Book. Available at: <www.icar.org.in/contract_%25%2020farming/index.%20 htm>.

Jain Irrigation Systems Limited, More crop per drop. PowerPoint Presentation, January 2011, p. 24 (accessed March 2011).

JISL, 2008. Improved cultivation of white dehydrator onion. <http://www.jains.com/PDF/crop/cultivation%20of%20white%20onion.pdf> (accessed in July 2013).

NSSO, 2006. Some aspects of operational land holdings in India, 2002–03. Report No. 492(59/18.1/3), National Sample Survey Organisation, Ministry of Statistics & Programme Implementation, Government of India.

World Bank, 2013. World Development Indicators. <http://data.worldbank.org/data-catalog/world-development-indicators> (accessed in August 2012).

CHAPTER 10

Private-Sector Participation in Agricultural Extension for Safflower Farming in India: The Case of Marico Limited

G.D.S Kumar, S.V. Ramana Rao and K.S. Varaprasad
Directorate of Oilseeds Research, Rajendranagar, Telangana State, India

Contents

INTRODUCTION

Indian agriculture is essentially small farm agriculture, with the majority of farmers owning less than 1 ha of land. The number of smallholders has increased as the average farm size has declined over recent years. Small and marginal farmers make up over 80% of farming households in India. Extension has been provided primarily by the public sector in India, implemented through each state Department of Agriculture (DoA). Extension is organized differently in each state, with wide diversity in personnel numbers and program focus.

Knowledge Driven Development.
DOI: http://dx.doi.org/10.1016/B978-0-12-802231-3.00010-3

The extension staff of the DoAs operate at the district and block levels, which are administrative subdivisions—a block is a subdivision of a district. The quantity and quality of extension staff vary greatly across the country, but it is well known that staff numbers are low. The ratio of staff to farmers varies widely across the country, as does the capacity of frontline extension staff—for example, only 20% of staff are university graduates. Information is transmitted from the district and block extension staff to the village levels through contact farmers or para-extension workers (Figure 10.1).

Agricultural extension in India is at crossroads as the DoAs, the main extension agency, struggles to find a fresh approach after the end of the training and visit system. The shortcomings of public-sector extension provision in India are very well documented (Farrington et al., 1998; Indian Council of Agricultural Research, 1998). The major emphasis of extension is on the dissemination of production technologies. However, the needs of farmers and rural families often go well beyond such technology. The development of technology and subsequent transfer of technology to farmers are performed by two separate organizations. These organizations have very tightly defined and mutually exclusive roles, often leading to the work being performed in two silos with no coordination. Coordination and linkages between ICAR and SAUs and the state DoA for extension activities have been documented to be weak (Indian Council of Agricultural Research, 1998). Despite many reforms in agricultural extension in India in the past decades, the coverage of, access to, and quality of information provided to marginalized and poor farmers is still uneven, even after the implementation of the Agriculture Technology Management Agency (ATMA) model. Although ATMA has been highlighted as an innovative example of agricultural extension (Singh and Swanson, 2006; Swanson et al., 2008), it has failed to make any significant improvement in the way extension is funded and implemented (Sulaiman and Hall, 2008; AFC, 2010). This case study explores the initiatives of a private company to provide agricultural extension services to farmers in India.

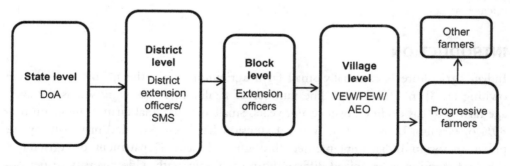

Figure 10.1 Information flow in the extension activities of the state DoA. *Note:* SMS, subject matter specialist; VEW, village extension workers; PEW, para-extension workers; AEO, assistant extension officer. *Source: Authors' compilation.*

This chapter also examines the replicability of the process in other value chains as well as key lessons that transcend other value chains for the delivery of agricultural extension services.

DESCRIPTION OF THE SAFFLOWER VALUE CHAIN IN INDIA

Safflower (*Carthamus tinctorius* L.) is a multipurpose crop with worldwide adaptability and greatly unexploited potential. It is one of the important *rabi* oilseed crops of India, cultivated in vertisols under residual soil moisture in Karnataka, Andhra Pradesh, Chhattisgarh, Madhya Pradesh, and Bihar. Safflower is known for its cultivation and use for orange red dye (carthamin), extracted from its brilliantly colored florets, and for its highly valued oil. The crop has superior adaptability to dry conditions. It produces oil that is rich in polyunsaturated fatty acids (linoleic acid 78%), which are effective in reducing human blood cholesterol levels. Dried red or orange flowers are often sold as a substitute for saffron in the markets of the Middle East and are used to color foods and beverages. Safflower petals are used for preparation of tea and in a number of medicinal preparations. Oil is used in the preparation of margarine, mayonnaise, and salad dressings and in the manufacture of paints, varnishes, and linoleum. Undecorticated safflower cake is generally used as a manure substitute and decorticated safflower cake is used as cattle feed.

India is one of the largest producers of safflower in the world, with an area of 170,000 ha and production of 97,000 tons in 2012–2013. Safflower seeds are primarily used for the production of oil for domestic consumption in India. A large proportion of seed produced is also retained by farmers for household consumption. The safflower value chain in India revolves around safflower producers and the processing industry. At the production level, support for farm input supply, support for extension services is essential to provide farm management training, adaptive research, development of improved safflower technologies, and coordination and management of farmers' groups/commodity groups. The extent to which safflower farmers receive and take advantage of these services is a key determinant of their level of safflower production and yield. Figure 10.2 illustrates the different players and relationships in the Indian safflower value chain.

Traditionally, the safflower plants are harvested when the leaves and most of the bracteoles except a few formed flower heads become dry and brown. The plants are cut with the help of sickles at the base or uprooted manually or mechanically, and stacked in the field in the form of small and well pressed heaps. In recent times, combined harvesters and threshers have been developed; these have significantly greater threshing efficiency and also cause much less damage to the seeds. The oil content in the seed is most important. The oil is extracted either by subjecting the seed to cold dry pressure in a *ghani* (traditional method) or by mini-expellers, distillation, or solvent extraction (modern method). The extracted oil is refined, packed, and marketed to consumers.

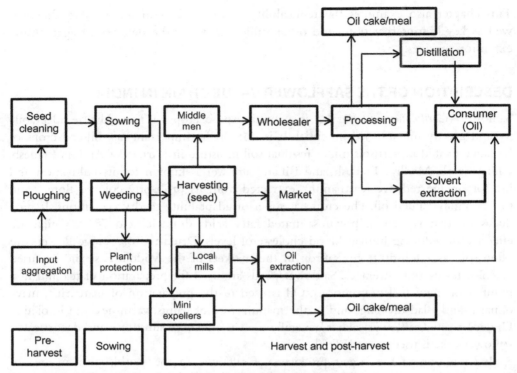

Figure 10.2 Safflower value chain in India. *Source: Author's compilation.*

At the farm production and farm processing levels, the major actor is the safflower farmer. However, in some locations safflower aggregators purchase seed from several small farms and processed at a central location. These buyers may or may not have their own farms, but usually have the resources to buy the safflower seed from the small producers. The most important knowledge required at the farm level includes updated information on improved safflower production technology. This is usually obtained through family apprenticeships, experience gained over the years, and through training organized by the extension departments and research facilities.

In recent times, the use of petals has been increasing, especially the petals of non-spiny cultivars, which facilitate easy plucking of petals. These new varieties were developed under the All India Coordinated Research Project (AICRP) on safflower. The popularization of these cultivars and their production technologies are the key to producing the required petals.

Solvent-extracted safflower meal in India is used primarily by cattle feed manufacturers, whereas expeller-pressed meal is used directly by farmers for feeding cattle. Safflower meal has moderate protein content (21–22%) but is very high in fibre

(35–40%), which discounts its value relative to other oilseeds' meal. High fiber also prevents its usage in the poultry industry (Nitin Kathuria, 2008).

In linking oilseed cultivators to the oilseed value chain in an economically viable and sustainable manner, the role of oilseed processing units cannot be underestimated. Past market intervention policies have not only failed to produce the desired results; they have also created an atmosphere of uncertainty in the oilseed market. The absence of a cohesive policy on market interventions has hurt the interests of oilseed cultivators and traders alike.

Oilseed crushing units include units in both the formal and informal sectors. The capacity of these units generally ranges from an average of 10% for *ghanis* (informal sector) to around 30% for expellers in the formal sector. The oilseed companies are currently functioning far below their capacity, mainly due to the low level of supply of raw material as well as poor organizational management. While there is excess processing capacity, production costs (e.g., cost of raw material, cost of money, cost of power, and logistics) continue to rise. In addition, imports of safflower have lowered the competitiveness of the Indian industry (Table 10.1).

With the implementation of FSSAI Act, the edible oil industries are now governed by FSSAI and must comply with strict license, safety, and standards parameters. However, the data monitoring of procurement for the edible oil industries are being administered by DVVOF under the Vegetable Oil Products, Production and Availability (Regulation) Order of 2010.

Table 10.1 Status of the vegetable oil industry (estimates as of January 2012)

Type of vegetable oil industry	No. of units	Annual capacity (100,000 MT)	Average capacity utilization
Oilseed crushing units	150,000 (approximately)	425 (in terms of seeds)	10–30%
Solvent extraction units	795	419 (in terms of oil-bearing material)	34%
Refineries attached with Vanaspati units	127	51 (in terms of oil)	45%
Refineries attached with solvent units	226	37 (in terms of oil)	29%
Independent refineries	590	35 (in terms of oil)	36%
Total refineries	943	123 (in terms of oil)	37%
Vanaspati units	268	58 (in terms of Vanaspati, bakery shortening, and margarine)	19%

Source: Directorate of Vanaspati, Vegetable Oil and Fats.

DESCRIPTION OF THE EXTENSION APPROACH

The current extension system is focused on a few major crops of the respective states. Safflower, due to its declining trend in area and production, is not considered a major oilseed crop and hence receives little funding for technology transfer by the DoAs. The lack of commodity-specific extension for crops such as safflower is the major constraint in effective technology and information dissemination. The public extension system is very limited in its ability to distribute subsidized inputs to progressive farmers wishing to grow safflower. Smallholder farmers, for whom the extension services are intended, are rarely able to generate enough support for improved extension services at the state or central levels. Despite the weak and uncoordinated extension services, safflower has the potential to contribute substantially to the oilseed industry with appropriate technology and training.

As farming is becoming increasingly knowledge intensive, farmers need support to access and use the available knowledge effectively. Since the safflower value chain is being neglected or given limited attention by the public system, private-sector operators are starting to provide extension services and purchase raw materials from the farmers. The provision of private extension service, however, is not well understood or documented. Although private funding in extension is desirable for specific commodities, public support will be needed to ensure extension services for farmers growing other food and commercial crops in a sustainable and equitable way.

In India, Marico Limited is one of the private sector companies working in the safflower value chain to provide extension to farmers. It is one of India's leading consumer product and services companies operating in the beauty and wellness industry. The aim of this case study is to provide an example of what the private sector is doing in regard to provision of extension services in India and what lessons can be learned.

Marico was established in 1990 and is currently the largest buyer of safflower in India. Saffola is a health care brand of Marico. The company's portfolio spans product categories such as edible oils, functional foods, and salt. Since risk factors for cardiovascular diseases (CVDs) start early in India and prevention of these diseases requires an integrated life course approach, Saffola has built a system which works actively toward adopting a healthier lifestyle. For over two decades, Saffola has played a pioneering role in leading the cause of generating awareness and motivation for lifestyle changes. This is among the most critical health concerns in India, with more people dying of coronary heart disease than any other cause. Over the years, Saffola Healthy Heart Foundation has taken a lead role in bringing together all the stakeholders in this cause—media, government, NGOs, hospitals, doctors, private health service providers, and citizen groups—perhaps among the rare occasions where such synergy has been achieved across such a diverse set of stakeholders. Over the years, Saffola has reached out to millions across the country via large-scale mass media awareness programs, helped over 100,000 people in 90 cities with diagnostic check-ups, and pioneered First Dial, a dietician service, and the "Heart Age Finder" in India. Currently, Saffola is among the most honored and rewarded brands in the country—by consumers and the industry. This has resulted in continuously rising demand for premium safflower oil in India.

Extension program content, target groups, and partners

The major emphases of the Marico Limited extension and advisory services for safflower farmers are presented in Table 10.2. The program is designed to target safflower farmers and assist them with the following:

1. promoting safflower as a very strong and viable alternative option in the *rabi* season by adopting a win–win approach for farmers;
2. providing need-based advice to farmers on improved technologies for quality safflower seed production; and
3. arranging for buyback of the produce, ensuring a regular supply of safflower to the company.

The private extension by Marico depends heavily on other public sector–supported programs. The primary partners in this program are the Directorate of Oilseeds Research (DOR) and AICRP on safflower centers supplying quality and timely inputs to farmers. In addition, the government of Maharashtra is responsible for mobilizing and reaching to large numbers of farmers. The Department of Agriculture and Cooperation (DAC)

Table 10.2 Areas of focus for Marico Limited

Problem	Emphasis		
	Not at all	Minor focus	Major focus
Closing technology gaps			√
Pest control			√
Natural resource management			√
Closing management gaps		√	
Providing market and input information			√
Output marketing			√
Input supply			
a. Seeds			√
b. Fertilizer			
c. Crop protection products			
d. Equipment/machinery			
e. Other			
Credit (leasing, payment structures)	√		
Distribute subsidized inputs to eligible beneficiaries	√		
Collecting crop and administrative data and providing associated reports		√	
Quality assurance			√

Source: Author's compilation.

Table 10.3 Partners and their main roles in the Marico extension program

Partners	Main roles
Central public sector: Department of Agriculture and Cooperation (DAC)	Funding agency for conducting frontline demonstrations under Integrated Scheme on Oilseeds, Pulses, Oil Palm and Maize (ISOPOM)
Local government: government of Maharashtra	Agriculture department of the state to provide awareness and farmer mobilization
Research institutions: DOR and AICRP on safflower centers	Technology backstopping and quality seed of improved varieties; handbook on safflower production
Farmer organizations	Mobilizing farmers to organize themselves into safflower farmer groups

Source: Author's compilation.

under the Ministry of Agriculture help to conduct demonstrations to show the productivity potential and profitability of improved technologies for safflower. Marico identifies progressive farmers, trains them, and supplies quality seed of improved cultivars on payment with the help of AICRP centers and DOR. The quality of the produce is ensured by frequent visits to farmers' fields by Marico staff, scientists of AICRP centers, and DOR. Table 10.3 illustrates the many key actors in the safflower value chain and their roles in the extension system.

Extension program approach and design

The program was designed to close technology gaps at the farm level and to ensure quality safflower seed production. Marico launched "Farmer First" in safflower-growing belts in June 2012, with the vision of achieving socially responsible growth by keeping farmers as the pivot. Marico has entered into a public–private partnership arrangement with the government of Maharashtra under which the company has covered 1,250 acres of safflower area and 575 small and marginal farmers. In this arrangement farmers were given all inputs (i.e., seeds, fertilizers, and pesticides) free of cost and also given assurance of guaranteed sale of their produce. Marico also regularly works with prominent seed companies and research agencies to ensure availability of high-yielding seed varieties to farmers at competitive costs (Table 10.4).

The AICRP on safflower consists of around 12 centers located in major safflower-growing states of India. Marico partnered with DOR to combat the declining trend in safflower area by showing the productivity potential in farmers' fields by adopting GAP. This program was able to maximize farmers' profits by keeping the farmer at the center of the process. The main source of technologies and approaches in the program is DOR and its network of AICRP safflower centers. DOR, a government-funded research institute, provides seed of improved cultivars and the production and protection technologies for conducting frontline demonstrations. The institute also participates in capacity-building activities for farmers and extension personnel.

Table 10.4 Extension approaches used to reach farmers

Type of approach	Organized by	Major focus	Scale and frequency
Farmer training	AICRP centers and DOR for progressive farmers	Good agricultural practices (GAPs) on safflower	At least one training session at each location during *rabi*
Demonstration sites	Progressive farmers	Showing the potential of improved seeds, production, and protection technologies	75 demonstrations during *rabi*
Field days (harvest)	Marico and government of Maharashtra and DOR	Interaction of farmers, scientists, and agriculture department people, and checking compliance with GAP	One at each demonstration site
Field visits	Marico, AICRP centers, and DOR	To assess crop performance and provide advisory services	One to two visits during the crop season
Provision of crop guide or pamphlet	Learning material in local language developed by AICRP centers and DOR	Improved production and protection technologies	Selected participating farmers—1,000 handbooks; other farmers; pamphlets
Arranging exposure visits to best-practicing farmers	A few selected farmers visited the DOR research farm	GAP on safflower	Once a year
Free input and supervision for new farmers	Marico	Recently released improved varieties	575 small and marginal farmers; free inputs; 75 demonstration farmers
Farmer innovators	DOR	Rewarding two to three innovative farmers	Once a year
All India Radio platform	AICRP center, Akola	GAP on safflower	18 districts of Maharashtra

Source: Author's compilation.

Marico joined hands with an AICRP center, Parbhani, to develop a handbook on the recommended package of practices in safflower production and distributed these booklets among 1,000 farmers across Maharashtra. Marico also worked with AICRP Akola to develop a farmer-friendly solution to the gujhia weevil, a major pest in Vidarbha region that destroys large quantity of safflower crop. Subsequently, an All India Radio platform was used to telecast the recommended package of practices across 18 districts of Maharashtra. Marico has engaged 6,632 farmers for contract farming of safflower seeds over 26,000 acres. Marico provides information and support throughout the crop cycle to assist farmers in increasing their yields. Marico also provides risk-free assurance for prices and yields. The company encourages farmers to adopt the practice of briquetting, which helps improve farm income and also provides a greener fuel.

Marico has established a well-defined quality control system. The company has engaged approximately 50 field staff, who monitor the crop throughout the season, providing need-based advisory services to the farmers. The company prefers safflower seed with high oil content. In order to achieve this, the company advises the farmers on GAP. Farmers must document the management practices they choose to adopt, which are then checked by the company's field staff. The field staff visits each farm to ensure that the farmers have followed the guidelines given. Once inspection by the field staff is completed, the staff of AICRP/DOR/Marico visits the fields to ensure quality seed production. Formal certification is not performed on the farm produce, but the visits of the field and technical staff from research and company ensure quality seed production. At the processing level, Marico adopts stringent environment management standards as governed by ISO 14000. The quality control system ensures a better price to the farmers and a regular supply of quality raw inputs to the company. Farms are checked regularly, and the field staff of Marico and Marico's program managers are accountable for improvements in the productivity and quality of the produce of the participating farmers. Once the collected produce reaches the center, quality checks and payment are completed on the spot and in front of the farmer to ensure transparency.

The overarching objective of the extension program is to revitalize safflower farming, make safflower a remunerative alternate crop during *rabi* season, and make it attractive for more youth to actively participate in its cultivation. This is partly a social corporate responsibility action on the part of the company, but in the long run it should bring higher profit for all the actors in the value chain, as well as a positive corporate image for the company. The reward for the farmers is mainly in the form of an assured premium price (10% over the existing market price) for their production and timely advice for crop production.

Cost sharing and cost recovery

The company bears the total cost of inputs such as seed, fertilizer, and pesticides for small and marginal farmers (~575 farmers). A small portion of the program costs are covered in part (demonstrations) by a grant from ISOPOM (around 75 farmers) and the major part by the company. In addition, other participating farmers (about 1,000) pay directly for the inputs. The input dealers sometimes supply inputs on credit to farmers, who repay the dealers when they sell their products.

ANALYSIS OF THE PRIVATE EXTENSION PROGRAM

Perspectives of the producers

Farmers and traders claimed that because of the myriad problems confronting the oil-seed sector, youth are not attracted to the business. Most safflower farmers are aging, and unless appropriate action is taken, safflower production in the country will be adversely affected. There is a need to link youth to safflower production, processing, and the private extension system. Identifying young farm leaders as progressive farmers in the program is one way to attract youth to safflower farming.

Several challenges confront smallholder safflower producers which the private extension system tries to reduce. According to many safflower farmers, markets do not offer remunerative prices for safflower seed, and hence they prefer to grow more profitable crops such as chickpea, maize, wheat, and other legumes. Although a minimum support price is announced for safflower seed by the government of India every year, procurement is not assured and the farmer has to depend completely on middlemen and regulated markets. Middlemen often cheat farmers with low prices and inaccurate weighing, and transportation to regulated markets creates additional costs to the farmers.

In most cases, inputs such as fertilizer are not available to farmers in a timely manner. Farmers prefer to use these limited fertilizers for more remunerative crops that are more popular at the market. Labor shortages during harvest are a severe constraint for safflower production. Mechanical harvesting is preferred by the farmers, but it is costly for them to acquire the harvester. If farmers were able to own small harvesters, this would help ensure timely harvesting and greater quality at the farm gate level. Thus, improving the quality and quantity of safflower production among smallholders crucially depends on improving extension and advisory services.

At the moment, private extension services are tailored to meet the quality requirements of safflower seed production and processing. However, farm-level innovations have not yet been incorporated into extension services. Including these developments in the extension system would benefit other farmers and help increase efficiency in reaching program targets. For example, in order to earn additional income, some farmers have improved indigenous technologies to add value to safflower by-products by preparing herbal health tea from dried florets of safflower. Another area where extension services could be improved relates to the harvesting of petals. As safflower is a spiny crop, harvesting petals is difficult and labor intensive. Petal harvesters were developed by the research system, and this advancement needs to be popularized among farmers through extension.

Public–private partnerships could aid in the development of the private extension system. The farmers believe that government (or other stakeholder) intervention is needed to effectively scale up these technologies.

According to some farmers in the state of Andhra Pradesh, it is difficult to access the seeds of improved safflower varieties. Farmers reported that when they attempt to procure the improved varieties from the researchers, limited quantities of seed are available to them. This is another opportunity for private extension to further help by supplying seeds of improved safflower cultivars.

Chudaman Patil, a safflower farmer from Maharashtra explains, "Marico's contract farming was a boon to us! We got rate guarantee and technical guidance throughout the crop cycle."

The farmers who participate in the Marico extension program explained that despite the achievements of the program in improving the yield and quality of saf-flower seed, one element still significantly threatens the ability of farmers to produce safflower that meets Marico quality criteria—the changing climate. Changing climatic conditions, such as severe drought during *kharif* and untimely rains in *rabi*, prevent the farmers from producing high-quality seed. This has adversely affected the entire saf-flower value chain. Developing innovations for adapting to changing weather patterns would help farmers meet the production and processing targets.

Farmers and traders claim that government policies have been unfriendly to the oilseed industry in general and safflower in particular. Up to 1992, India was nearly self-sufficient in terms of edible oil consumption. In 1994, edible oil imports were brought under Open General License (OGL), and from 1994 to 1999 the rate of cus-tom duty was the same for crude and refined oils. In 1999, the Indian government reduced the duty on import of crude oil to encourage value addition of refining within the country, which made cultivation of oilseeds less lucrative for farmers.

Perspectives of the extension workers

Marico extension officers found that young farmers are getting involved in safflower production under the extension program established by the company. It was found that the young farmers were motivated by the agribusiness focus of the program and the visible link between production in their respective villages and the premium in the domestic market for safflower oil and petals. In addition, the extension officers found that the younger farmers are eager to take advantage of the available training. Most farmers could see a direct link between the training, farm productivity, and the com-petitive advantage they gain from adopting the lessons from the training. Business ori-entation, use of information, use of communication technology, and providing access to buyback arrangements are the key drivers of young farmers' interest in safflower production and processing. Private extension systems should focus on these aspects in order to attract younger farmers to safflower production.

The extension officers also noted that farmers face the most challenges during peak harvesting season, due to labor shortages coupled with the unwillingness of farm laborers to work in safflower fields due to the spiny nature of the safflower plant. Alternatively, mechanical harvesting of the capsules must be encouraged to overcome these labor problems. The program does not currently conduct any research or innovation in this area, although it is recommended to be explored. At present, the DOR is the program's primary research partner. Resources for the DOR and other relevant institutions (such as universities) to research alternate harvesting methods are needed from the public sector in order for the private-sector extension in safflower to be sustainable and innovation oriented.

Finally, improving rural infrastructure and increasing the education level of farmers would contribute to sustainable development in the sector. For example, the current extension program does not provide support in some key areas, such as farm-to-market transportation. Farmers resort to selling the produce to middlemen, who come directly to their fields. A reliable road network would increase safflower farmers' market access and enable them to get better prices for their produce. It is also suggested that a regular farmer training and retraining program be established. Currently, Marico does not have the resources to develop safflower educational programs for the communities it serves.

Perspectives of the company

In India, extension services are provided mostly by publicly funded agencies. Commodity-specific private extension is offered on a limited scale, usually by the companies that engage the farmers for selling inputs and/or buying the output. Marico started this program in order to enhance the livelihood of smallholder safflower farmers through effective linkages with research and an assured market.

In the context of Marico, the process begins with the training of progressive farmers (~50), who are expected to share the knowledge they gain with other safflower farmers. The training program concentrates on GAP, GEP, and GSP for safflower production. DOR and AICRP safflower centers train the progressive farmers and provide technical support for field advisors. The centers also facilitate the supply of quality seeds to farmers. The current varieties grown by farmers have around 25% oil content, whereas the improved varieties (PBNS 12, NARI 38, and SSF-708) have shown higher yields and about 30% oil content. The key performance indicator is the number of quintals of safflower seed sold to the company. Thus the major focus of the private extension program provided by Marico is to increase the area under improved varieties.

As a matter of courtesy, when the company works in a state, the state Agriculture Department, the primary functionary that is supposed to deliver advisory service at the state level, is informed of the company's activities. In general, there is a high level of interest within Marico in increasing the safflower market in India, including partnerships for research and extension. Marico extension program with the contract farming model have led to expansion in acreage planted due to the company's buying price guarantee, accurate electronic weighing, same-day cash payment, and technical guidance throughout the crop cycle.

Perspectives of the community at large

At one time India produced more safflower than any other country in the world. However, India's share of world safflower area and production has been steadily declining for the past two decades. This decline has been attributed to many causes. The principal reason for the decline has been the increased profitability of competing crops such as sorghum, thus making safflower production less attractive to farmers. The average price obtained for safflower between 1991 and 2003 was less than Rs. 12,000/ton, lower than for any other oilseed. In addition, the low price of imported palm oil has also deterred domestic safflower production in India. All these factors have resulted in a continuous decline of safflower area in the country, from more than 1 million ha in 1994–1995 to 170,000 ha in 2012–2013.

The safflower processing industry is integral to the oilseed sector. A vibrant and efficient processing sector is crucial to the growth and development of the safflower market. India's oilseed processing sector has been plagued by a slew of technological and policy issues leading to low levels of efficiency and capacity utilization. Thus, provision of extension by the private sector could increase the quantity and quality of the safflower produced, leading to improved capacity utilization and efficiency through continuous supply of raw material.

The executive director of the Solvent Extractors Association of India identified the following challenges confronting processors in India:

- The refining margins for the industry are zero to negative, and the industry is in a deep crisis and on the verge of closure;
- Low capacity utilization;
- Lack of raw materials and their continuous flow for processing;
- Insufficient financing/credit finance;
- High cost of machinery and spare parts;
- The domestic vegetable oil industry had demanded a rise in the custom duty on imports of oils to create a duty difference between imports and domestic product. The government of India partially acceded, announcing a rise in the import duty on refined oils to 10% from 7.5%, creating a difference of 7.5% between crude and refined to protect the domestic refining industry and domestic farmers.

Overall, all the stakeholders agree that the Marico advisory service is efficient and effective in meeting the needs of the company. The extension ratio of 1:75 (progressive farmer to farmer), as compared to 1:1,000 or worse in the government-run extension service, is considered appropriate for the program's target. This ratio is easily manageable in tailoring production recommendations to meet the needs of individual farms and rural households. The sustainability of the Marico advisory service is currently hinged on its cost-recovery approach, discussed above. So long as the safflower processing industries and related businesses continue to run at a profit, the incentive to continue the service will remain high.

COMPARISON WITH PUBLIC EXTENSION AND OTHER PRIVATE SECTORS IN INDIA

The primary motivation of the private sector to invest in extension services in India is to enhance the quality and assurance of continuous supply of safflower seed to ensure profitability in processing. Among other factors, the inefficiency of government interventions in the oilseed sector, and safflower in particular, which negatively affected the processing industry, and changes in domestic consumption of safflower oil have ultimately led to the emergence of private extension services. Although the pace of private-sector investment in extension services has been slow, it is seen by farmers as a welcome complement to the government-run extension programs. In addition to securing the supply of raw materials needed for their operations, the idea of corporate social responsibility is gradually extending beyond the immediate profit motives of the processing companies.

It is important to note that farmers continue to request government involvement in the funding and delivery of agricultural extension services. This is mainly due to the fact that government services are free and that government providers could fill gaps that the private sector cannot deliver on its own. However, the opportunities created by public funding and management of extension services are not fully lost. Many farmers interviewed advocate that, at the very least, the government must provide supportive and regulatory services to prevent exploitation by privately run extension services. Farmers also report that the resources currently spent by the government on agricultural and rural development programs could instead be channeled into financial institutions. These institutions could provide relatively cheap investment resources for young entrepreneurs in both the safflower industry and other agriculture enterprises. Thus, it is safe to say at this point that private extension services in India currently complement public extension services, but do not fully replace them. The implication of this is that public extension services need to evolve to optimize the resources that government and the private sector invest in agricultural extension services. The model of extension services adopted by Marico can be followed for other commercial commodities where companies can profit from increased farmer productivity.

The key value addition in the Marico model is that the motivation to produce quality seed due to premium buyback arrangements for farmers offers a simple model for sustainably maintaining profitable smallholder safflower farms. Marico is currently offering advisory services whose costs are mainly borne by the company as a form of corporate social responsibility. Although farmers do not directly pay for information received, the current system is highly cost-effective because it provided benefits for both the farmers and the company. This is mainly because the adoption of farm technologies and advice is directly related to farm outputs and therefore increased income for participating farmers. However, in the public extension service this relationship is not as clear due to the wide range of services that the public-sector extension is required to deliver. Further, the public extension service is primarily focused on food security and therefore continues to promote technologies that encourage farmers to produce more food. This focus of the public extension service, along with the perception of safflower farming as an unremunerative enterprise have resulted in limited public extension services for safflower farming.

UNIQUENESS OF THE MODEL AND VALUE ADDITION

The Marico Safflower Extension Program is a new service developed primarily to meet the company's industrial needs for a regular supply of high-quality raw materials that satisfy standards for high oil content. It therefore aims to deliver an extension service that increases safflower productivity and quality, and through this pathway to raise farmers' income. There are three unique features of this program:

1. It makes the profitability of safflower production for farmers a key consideration in its operation rather than simply focusing on increasing levels of safflower production for its own supply. The economic returns to the farmer provide an incentive for them to continue to produce following best practices, as recommended by the private extension. It also makes safflower production more attractive to youth who are not much interested in safflower farming as an agribusiness rather than as a way of life.

2. The drive to generate a high-quality product to meet the requirements of premium oil for domestic consumption has the additional advantage of making safflower farming more competitive in India. The Marico brand, Saffola, attracts a premium over current oils and is more attractive to a wider range of buyers. Therefore, the production of high-quality safflower is driven both by the expectation of higher economic return and by the opportunity to explore new markets. The extension program of Marico has been able to bring the actors and players together to achieve these goals.

3. Direct linkages between producers (farmers) and processors (industry) of safflower seed tightens the supply chain, leading to more efficient communication and coordination between actors. This also helps the extension service to remain demand driven and efficient, focused on issues highly relevant for smallholder safflower producers. Although this feature increases the impact of the extension service for a specific value chain, private extension requires a lot of time from the farmer. This may come at a high cost, especially if the farmer produces multiple crops and/or

raises livestock. Conversely, in the public extension service, a single extension agent exchanges information on different crops and livestock with farmers in a holistic manner, though the quality of service may be low and have inadequate coverage. The implication of this is that in the long run either smallholders will become more commercially oriented and concentrate their efforts on fewer crops and/or livestock, or the nature of the private extension services will need to evolve in ways that meet the diverse needs of small farmers rather than those of the company that promotes it.

LESSONS FOR REPLICATION AND SCALING UP

The Marico extension program has been in operation since 2012, reaching more than 6,600 farmers by the end of 2013. It plans to scale up the program from its current operations in Maharashtra and expand to other states in southern India.

The main challenge to sustaining the program currently is respect for the contractual arrangement. For example, the incidence of side-selling of safflower seed after reaching an agreement with the company remains high. Under the current contracts, farmers have the choice of side-selling but will have to pay for inputs already acquired through the program up to the time of exit. In order to overcome this problem, the contracts need to be modified to prevent side-selling.

Despite these challenges, the Marico extension program offers a model that can be replicated by other companies. The model is easily replicable because it is simple and directly rewards farmers according to the quantity and quality of safflower seed sold to the company. It improves farmers' income as compared to selling the seed in the market or to dealers. The Marico extension program aims to increase self-reliance of smallholder safflower farmers as opposed to total dependency on external actors. This means treating smallholders and other value chain stakeholders as development partners.

Integrating issues of equity, self-reliance, and sustainability on a large scale is essential, although challenging.

The Marico extension program has taken these factors into account by anticipating potential tensions stemming from the rapid development of safflower value chains, the goal of aiding smallholders, and addressing sustainability. Long-term planning and an adaptive problem-solving approach have been used to build capacity along the entire value chain and align the key production and processing elements of the safflower value chains.

Furthermore, the individual and organizational capacities of smallholder safflower farmers are strengthened as they engage in more commercially oriented safflower farming. Farmers with fewer resources, who require significant support in order to benefit from the safflower value chain development, are a special target of the Marico extension program. Strong farmer organizations allow individual smallholders and processors to benefit from the value chain through collective action. The Marico extension program uses a cost-effective approach to strengthening individual capacity, and uses the potential of different types of farmer organizations. The program works with partners from the public, private, and NGO sectors to utilize their ability to secure resources and provide sustainable services. For example, the Marico Foundation has educated consumers on the benefits of using safflower oil in reducing the risk of CVDs and thereby has increased the demand for safflower oil.

Finally, the Marico extension program draws on the experiences of its staff, including the various relationships and networks they have cultivated over the years. These positive strengths enable the company to scale up their model. Developing a new, inclusive smallholder value chain requires further investments in developing and nurturing value chain relationships and aligning key actors and elements. Building strong relationships and networks along the value chain creates trust, increases understanding of interests, and clarifies expectations. The greater the social difference between value chain actors, the greater the investment needed in relationship building.

CONCLUSIONS

Private extension systems are increasingly able to complement the functioning of public extension systems. Safflower production among smallholders in India and the extension provided by the processing companies are increasing. This case study described the issues, constraints, challenges, and possible improvements to private extension provision for safflower production.

India was one of the largest safflower growing countries in the late 1980s, contributing up to 69% of the world's safflower seed production. Though a twofold increase in the productivity of safflower was witnessed in the last three decades, the present average productivity of 575 kg/ha is still very low compared to the production potential of improved safflower production technologies. The major reasons for low productivity of safflower are the lack of availability of quality seed, lack of awareness of improved production technologies, the incidence of pests (aphids), and unremunerative market prices for safflower seed. Apart from these constraints, the public extension system does not put much emphasis on safflower production. The crop is considered a minor oilseed and lacks commodity-specific extension efforts. However, with the increasing domestic demand for high-quality safflower oil and investments in safflower value chains by domestic private investors, there has been an increased role for the private sector in providing extension services. In the case of the Indian safflower industry, it was realized that the efforts of the public extension system were inadequate. However, safflower farming is a knowledge-intensive activity for which farmers need support to access and use available knowledge properly. As the public extension services are not adequately meeting the information needs of smallholder safflower farmers, private-sector operators are beginning to provide extension services. The provision of private extension services, however, is not well understood.

Marico is one of the private sector companies working in the safflower value chain, providing extension and advisory services to farmers. This case study has provided an example to help us understand private-sector extension from the perspective of a company. Marico provides these services in order to improve the productivity of the safflower producers, increase the quality of the safflower seeds to meet domestic demand for premium oil, improve the availability of inputs in a timely manner, and connect various actors and players in the value chain to ensure a regular supply of safflower seed for processing. Inadequate government policies toward the oilseed industry have pushed the safflower processing industry and smallholders alike to form an alliance to help each other in the form of private extension. The private sector is able to meet information and input needs of the farmers that are demand driven and relevant to the farmers' needs. The extension services provided are delivered effectively and in a timely manner through a system of progressive farmers who connect remaining farmers to the extension agents on a regular basis. Cost recovery through the price paid to farmers has been a win–win situation for both the farmers and the company, as the price farmers receive is higher than that paid by local buying agents. This model of extension is replicable and scalable with needed modification to other states and other production environments.

REFERENCES

AFC, 2010. Report on evaluation and impact assessment of the centrally sponsored scheme Support to State Extension Programs for Extension Reforms (ATMA) in Uttar Pradesh and Haryana, submitted to the Department of Agriculture and Cooperation, Ministry of Agriculture, Government of India. May 2010, Agricultural Finance Corporation of Limited.

Farrington, J., Sulaiman, V.R., Suresh, P., 1998. Improving the Effectiveness of Agricultural Research and Extension in India: An Analysis of Institutional and Socio-Economic Issues in Rainfed Areas. Policy Paper 8. National Centre for Agricultural Economics and Policy Research, New Delhi, India.

Indian Council of Agricultural Research, 1998. National Agricultural Technology Project: Main Document. ICAR, New Delhi, India.

Kathuria, N., 2008. Safflower markets in India. Seventh International Safflower Conference, 3–6 November, 2008. Australian National Wine and Grape Industry Training Centre, Wagga Wagga, NSW, Australia. <http://safflower.wsu.edu/Conf2008/Key%20Note/Nitin's%20Presentation%20Summary.pdf>.

Singh, K.M., Swanson, B.E., 2006. Developing market driven extension system in India. In: 22nd AIAEE Annual Conference, International Teamwork in Agricultural and Extension Education, May 14–19, 2006. Clearwater Beach, FL, pp. 627–637.

Sulaiman, V., Hall, A., 2008. The fallacy of universal solutions in extension: is ATMA the new T&V. Linklook, pp. 1–4.

Swanson, B., Singh, K.M., Reddy, M.N., 2008. A decentralized, participatory, market-driven extension system: The ATMA model in India. Paper Presented at the International Food Policy Research Institute Conference "Advancing Agriculture in Developing Countries through Knowledge and Innovation," April 7–9, Addis Ababa, Ethiopia.

CHAPTER 11

Private Provision of Extension Through the Contract Farming Approach in Bangladesh: The Case of PRAN in Bangladesh

Fatema Wadud[1], Suresh Chandra Babu[2] and Safiul Islam Afrad[3]
[1]Directorate of Agricultural Marketing, Ministry of Agriculture, Dhaka, Bangladesh
[2]International Food Policy Research Institute (IFPRI), Washington, DC, USA
[3]Banga Bandhu Sheikh Mujibur Rahman Agricultural University, Gazipur, Bangladesh

Contents

INTRODUCTION

The agricultural sector is central to Bangladesh's economy. About 80% of the population live in rural areas, and most of them depend on agriculture for survival (World Bank, 2012). According to the 2010 Bangladesh Bureau of Statistics household survey, 31.5% of

Knowledge Driven Development.
DOI: http://dx.doi.org/10.1016/B978-0-12-802231-3.00011-5

the population lives below the poverty line, and 17.6% are said to be "extremely poor." In recent years, however, the country's GDP growth rate has been above 5%. Although this pace of economic development is satisfactory, some development practitioners believe that it could have been much stronger (Anon, 2013). Despite various challenges, including economic slowdowns in the West, Bangladesh's export sector has gained momentum, which can be credited, in part, for the country's recent growth. Processed agricultural products make up a large proportion of Bangladesh's growing export sector. Both the number of items of exportable processed foods and the number of countries to which products are exported are increasing. In the last 2 years, the growth in the quantity of goods exported was approximately 15%. According to the Bangladesh Agro-Processors' Association, exports were worth US$52.28 million in 2009–2010 and US$59.15 million in 2010–2011 (BAPA, 2013).

Most of the country's agricultural products are not processed, despite the numerous benefits associated with processing yields. Agro-processing is considered an appropriate intervention in reducing post-harvest loss, generating fair prices for producers, generating employment, extending the shelf life of the commodities, and contributing to improved nutrition. Recent interest among Bangladeshi entrepreneurs in agro-processing led to a 22% increase in the contribution of this sector to national GDP. One unique opportunity for agro-processing is due to the fact that there are 8 million non resident Bangladeshis, creating great demand for imported Bangladeshi agro-processed commodities in the countries in which they now reside (Quddus and Mia, 2010). Some of the major private companies working in this sector include the Program for Rural Advancement Nationally (PRAN), Advanced Chemical Industries, Bangladesh Food (BD Food), and Square.

PRAN is the largest certified agro-food exporter in the agro-processing sector in Bangladesh. Its vision is to enrich the agriculture sector in Bangladesh. It looks both to create more demand for agricultural products and to enhance production by providing training and financial support to the participating farmers, many of whom are poor. PRAN wants to build its contract farming program to be among the largest in the country. In this way, PRAN aims to create more jobs and make its products available to every corner of Bangladesh so that every consumer has access to its products. In addition, it aims to bring more foreign currency into Bangladesh.

Performance of PRAN

In recognition of its contribution toward bringing in foreign currency, PRAN won the "Best National Export Award" for 8 consecutive fiscal years (from 1999–2000 to 2009–2010). PRAN was also awarded a 2011 UDC Business Award for the best food and beverage products manufacturer in Malaysia (PRAN, 2013). The company is increasing its exports, not only in terms of geographical area, but also in terms of US dollars. Figure 11.1 illustrates PRAN's span in terms of geographic area.

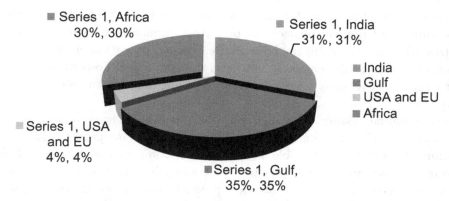

Figure 11.1 Proportion of PRAN export in different parts of the world. *Source: From PRAN (2011).*

The number of farmers contracting with PRAN has continued to grow across the country in recent years. There are 78,000 farmers under contract to PRAN, located in different areas of the country, namely Natore, Rajshahi, Chapai Nawabgonj, Dinajpur, Gaibandha, Lalmonirhat, Khagrachari, Panchagarh, Munshigonj, Shariatpur, Jessore, Satkhira, and Naogaon districts.

Most of the workers on the production floor are female, and there is a noticeable presence of female supervisors in the factory. The company provides new workers with technical and safety training. At least two workers are trained for each task so that there is always someone who can safely and efficiently replace another worker when necessary. Internships are available for students from engineering and technology institutions to give them an opportunity to gain practical experience. This is very encouraging for both women's empowerment and the enrichment of technical education.

PRAN is one of the most remarkable companies in Bangladesh, in terms of its commitment to corporate social responsibility. The company provides its workers with protective clothing, including an apron, cap, mask, and gloves, along with gum boots. The factory has supervisors who regularly check for appropriate use of this protective gear. There is a sufficient supply of chlorinated water to wash hands and foot baths to wash the workers' shoes so that they are clean before they enter the production floor. There are health and safety instructions posted on the factory wall in Bengali and, in some places, there are also instructional diagrams. With these measures in place, PRAN meets global standards for safety and security.

The company set up a youth club for the younger generation. It offers free coaching for primary and junior scholarship examinees and has started providing scholarships to their employees' children who demonstrate good performance at school. Plans have been made to begin gardening on selected school premises. There are donation programs to mosques and madrassas. The company encourages the spread of education and plans to establish a school for their employees' children. Currently, it provides the salaries for teachers in mathematics and English in selected schools. It contributes to the family and relatives of deceased or former employees who were disabled by industrial accidents. The company bears all the expenses of any burial and pays a good amount of money to the family. It also bears the educational costs of their children, and if a child or other close relative is able to do a job, the company hires them in the factory.

During natural disasters in Bangladesh, the company donates money and relief goods. PRAN's effluent treatment plant is modern and was designed to cause as little environmental pollution as possible. In addition, the company uses a variety of methods to increase demand for its products. For example, there is an annual cooking contest whose recipes are developed into new products. This approach has increased the demand for PRAN's processed commodities such as pickles.

The company has an extension strategy to enhance the agricultural products it buys as inputs for its processing factors, using a contract farming approach. Thus far, it has been observed that both parties have benefited from this approach. However, there is no documentation describing the model from which others can learn. Therefore, the present case study aims to:

- identify the extension service delivery approaches undertaken by PRAN;
- assess the quality of extension services provided; and
- analyze the success factors thus far and potential constraints in terms of long-term sustainability and development of a partnership model.

SWOT analysis of PRAN

Every organization has not only strengths and opportunities, but also struggles with weaknesses and threats. Strengths and weaknesses are internal, while opportunities and threats are external. PRAN's strengths currently include its high profit margin, its employees' strong devotion to the work, and its relatively high market share. Its opportunities stem from worldwide free trade agreements, mounting health consciousness in selecting foods, and increasing demand for branded foods. On the other hand, impending threats are uneven competition in the world market and increased demand for environmentally friendly containers and packaging (the company uses nonbiodegradable packaging). Table 11.1 shows a preliminary analysis of the effectiveness of PRAN.

DESCRIPTION OF VARIOUS VALUE CHAINS
Input supply and backward integration

PRAN follows a seasonal crop calendar, and so it takes care to ensure timely collection and distribution of seeds and other inputs through center heads. The center heads require the support of their subcenter head farmers for efficient distribution of the inputs. There are preset standards for different technologies and their distribution.

Table 11.1 PRAN's strengths, weaknesses, opportunities, and threats

	Strengths	*Weaknesses*
	1. Current profit ratio increased 2. Employee devotion to work 3. Market share has increased	1. Strategic management system
Opportunities 1. Worldwide free trade agreement 2. Rising health consciousness in selecting foods 3. Demand for brand food increasing annually	1. Develop new healthy food items 2. More expansion of company capacity	1. Develop new farm products
Threats 1. Uneven competition in the world market 2. Containers are not biodegradable 3. Unrest in sociopolitical situation of the country	1. Develop new biodegradable food containers	1. High-cost Western operations

Source: Author's compilation.

The center heads learn the standardized parameters from their block supervisors and receive these in printed form as well. PRAN obtains these parameters from both domestic and international research organizations. PRAN provides inputs at low cost to the farmers through its lead farmers but does not charge fees for rendering advice to the farmers.

Production advice and package of practices

Farmers always have support available to them under the company's intensive follow-up system with field-level personnel, and this has increased their loyalty to the company. The company's extension system provides advice on the selection of quality seeds, appropriate land preparation time, seed sowing, management practices, application of irrigation, harvesting, crop drying, grading, and so on. Regular, intensive field visits are carried out by block supervisors to encourage farmers to maintain their set standards during the crop season. The production package provided to farmers includes seed, fertilizer, pesticides, equipment, and recommendations for good management practices. Information sources for best management practices are obtained from research stations, where they are tested in the field. As soon as extension workers receive notice of a new problem in their farmers' fields, they inform the manager of the program to obtain timely solutions. The manager himself tries to find a solution and, if he cannot, he contracts with the local research station to do so. During the field day or farmers' day, local public extension workers are invited. Meetings are also arranged with them in agricultural hubs or local union parishad offices. PRAN works with its own farmers' organizations, where all the participant farmers are directly involved with the disseminating strategy under the control of their center head.

Harvest/post-harvest/storage guidelines

From its contracted farmers, PRAN purchases produce that meets the quality and standards laid out to the farmers earlier. Guidelines for safe storage are also passed on to the producers by the center heads and subheads, especially with respect to optimal moisture levels, protection from excessive heat and rain, and use of fumigation. PRAN has crop-specific requirements that producers must meet to ensure PRAN can meet its commitment to consumers at home and abroad. PRAN recovers credit provided to farmers by deducting the value of the loans from the final payment.

Marketing and contract arrangements

Usually, PRAN collects agro-products from the contract farmers and processes them in their factory. The processed food items are purchased by wholesalers, and retailers purchase from the wholesaler. Consumers get their PRAN branded food items from the retailers throughout the country and abroad. One of PRAN's more recent initiatives involves marketing some of its products, vegetables in particular, through a producer group, Swopno, a chain of superstores. Farmers' groups send vegetables that meet PRAN's criteria to the center hubs, and PRAN transports the products to the different Swopno locations each day. The main challenge is an insufficient supply of vegetables.

PRAN contracts farmers through its agricultural hubs. Contract farmers receive seeds, fertilizer, pesticides, irrigation equipment, recommended farm management practices, and credit if required by the farmers. Field-level extension workers determine demand for these inputs in each hub area. The costs of inputs and/or credit are recovered from the farmers when they sell their products to the company's hub. Center heads take responsibility for recovering credit from the farmers. There are some problems with side-selling or side-buying, and if this is noticed by company personnel, the farmers involved in this activity are blacklisted and barred from future dealings.

Terms of the contracts and prices are negotiated with the farmers during group meetings. Lead farmers at each hub bargain with the company on behalf of the farmers. When negotiating over the price of a product, a team is formed, consisting of representatives from the company and farmers as well as the center head and subcenter heads. They visit at least three nearby markets to determine local market prices. Then an average price is calculated for each product. Fluctuations in the price of a product is a regular phenomenon. When prices change sharply, the committee meets again to make the price comparable to the market price. There is no provision for profit-sharing among the parties.

An internal audit system monitors the contractual arrangements and company accounts, and the auditors sometimes facilitate negotiations between the parties. There is no government involvement in the negotiations, regulation of contracts, or their enforcement. Surprisingly, no conflicts or tensions between the farmers and contractors have been observed as yet, which increases the credibility of the company to the farmers. Farmers are found to be highly satisfied with their dealings with the company.

Quality assurance, standards, and certification

Block supervisors select and train lead farmers, but extension agents are selected and trained by program managers. Highly specialized subject matter specialists and program managers are trained by specialists at the research stations. There is a well-equipped quality control (QC) unit in the company. The lead farmers bring their products to the PRAN processing center gate. The QC unit for PRAN collects samples of each product and examines them in the laboratory to measure whether the samples meet the company's QC and safety standards for rejection or acceptance of the lot.

Focal crops are tomato, mung bean, peanut, mango, rice, chili, coriander, litchi, olive, turmeric, and tamarind. The contract farming and processing steps (value chain) for mung beans and peanuts are shown in Figures 11.2 and 11.3, respectively. After harvesting, farmers take their crops to the mill gate for QC. After receiving a positive signal from the QC unit, the products enter the processing system, where value is added at different stages of processing. The products are stored centrally, and then prepared for marketing by the wholesaler for domestic sales or for export.

DESCRIPTION OF PRAN'S EXTENSION PROGRAM
Program content

The foci of PRAN's contract farming program are: closing technology gaps; pest control; natural resource management; closing management gaps; providing market and input information; output marketing; input supply (seeds, fertilizers, crop protection products, and equipment/machinery); collecting crop and administrative data and providing associated reports; and certification and quality assurance. Minor foci of PRAN include credit (leasing, payment structures) and the distribution of subsidized inputs to eligible beneficiaries during production.

Program design

The key aim of the program is to support producers by ensuring fair prices for their products and minimizing the need for middlemen. PRAN AL has provided private extension services for selected farmers since 2001. Its approach is quite different from that of the government. Key features of the program are described in Table 11.2.

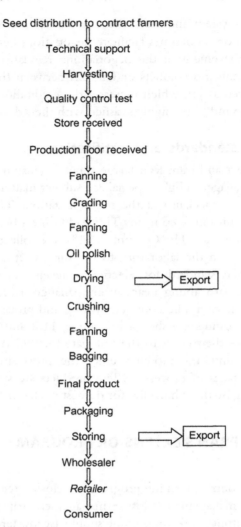

Figure 11.2 Value chain of mung beans.

When developing an extension strategy for a particular crop, PRAN extension personnel first identify the best growing area for each crop. Second, the names of interested farmers in the area are noted. In a preliminary meeting, the lead farmer is selected by the farmers according to criteria such as social acceptance, economic solvency, education, dedication to the group, and farm size. An agricultural hub is established as the center point for each area. The location of the hub is selected by the farmers themselves. During the meeting, in-depth discussions are held on a variety of issues, such as the objective of the meeting, the role of PRAN in improving farmers'

Seed distribution to contract farmers

⇩

Technical support

⇩

Harvesting

⇩

Quality control test

⇩

Store received

⇩

Production floor received

⇩

Crushing

⇩

Grading

⇩

Bagging

⇩

Final product

⇩

Packaging

⇩

Storing

⇩

Sending to wholesaler

⇩

Retailer

⇩

Consumer

Figure 11.3 Value chain of peanuts.

Table 11.2 Extension approaches and main features

Type of approach	Led by	Primary focus	Scale and frequency
Farmers' training	PRAN	Production technology	Broad scale and crop based
Demonstration sites (rice and tomato)	PRAN with Syngenta and Buyer Crop	Yield	Crop based
Field days (harvest)	PRAN and Department of Agricultural Extension	Performance	Broad scale and crop based
Field visits	PRAN	Production advice	Daily
Provision of crop guide or pamphlet	PRAN	Technical information	Crop wise
Arranging exposure visits to best-practicing farmers	PRAN	Motivation	Seasonal
Free input and supervision for new farmers	PRAN	Motivation	New agriculture hub
Joint training of farmers with other private/public actors	PRAN with Syngenta and Buyer Crop	Technical knowledge	Demand oriented

Source: Author's compilation.

socioeconomic status, the extent of farmers' involvement in the upcoming cropping scheme, the crops to be grown, and farmers' benefits from the cropping scheme. The following is a list of activities carried out when a new hub is established:

a. Crop-based area selection
b. Listing of farmers
c. Selection of center head
d. Selection of subcenter head
e. Site selection for hub
f. Training the farmers on crop production
g. Setting demonstration plot by Buyer Crop Company
h. Distribution of seeds
i. Distribution of credit (if required)
j. Visits and consultancies on management practices
k. Schedule spraying by Buyer Crop Company
l. Provision of irrigation equipment (if required)
m. Crop harvesting
n. Drying and grading
o. Collection of crops by the lead farmers
p. Transporting the crops to the PRAN center
q. Distribution of the checks to farmers
r. Payback of loan
s. Preparation for the next crop

The lead farmer is the key person in an agriculture hub. He organizes meetings under the instruction of the PRAN field supervisor. He alerts farmers to the appropriate time for the production activities including land preparation, collection of seeds, intercultural operations, installment of irrigation equipment, spraying of pesticides, harvesting of crops, grading and storing the harvested crops, and transporting the crops to the hub. Furthermore, the lead farmer makes all necessary arrangements to transport the harvested crops to the PRAN processing center. He has the responsibility of ensuring that all quality and safety standards set by PRAN are met. If the crops are taken to the PRAN processing center for delivery and they do not meet their criteria, the whole lot is sent back. Therefore, the lead farmer must ensure that the crops he collects are of sufficient quality. When the prices to be paid to the farmers are being determined, payment for the lead farmer for his supervisory duties and transportation responsibilities are also discussed. This structure is illustrated in Figure 11.4.

Once a lead farmer is selected, training sessions are organized for farmers on precision agriculture, that is, how to maximize yield given their land holdings. This involves arranging subject-specific training, such as in identification and collection of quality seed, optimum time for land preparation, intercultural operations, optimal dosages of fertilizer and pesticides, irrigation and its importance in crop production, and

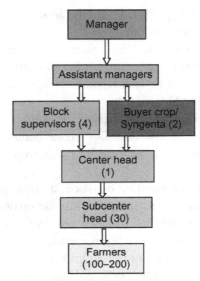

Figure 11.4 Organizational structure of an agriculture hub. *Source: Author's compilation.*

marketing. However, PRAN is responsible for marketing. Therefore, farmers do not have to worry about the sale of their produce to the consumer.

Each lead farmer or center head performs his/her work with the cooperation of several subcenter heads. The number of subcenter heads varies according to the size of the group. There is one subcenter head for every 100–200 farmers. Subcenter heads share responsibility with the lead farmers for recruiting farmers, motivating them, conveying information, monitoring the quality of their produce, and collecting harvested crops to transport to the hub. PRAN bears the costs of transporting the crops.

Buyer Crop Company and Syngenta are field-level partners of PRAN. They have a memorandum of understanding. Buyer Crop is responsible for crop protection and has field-level supervisors with diplomas in agriculture. They schedule sprays for different crops. They use bio-pesticides to promote sustainable agricultural production. They provide training for farmers on safe and clean agriculture.

The key innovation in this program is security for the farmers in terms of guaranteed sale (provided they meet the minimum standards for their products) and the elimination of middlemen. Computer facilities are available in the agricultural hub, and mobile phones and other ICT technologies are used in the program to disseminate information. PRAN extension agents receive their information from research stations and companies that supply inputs. PRAN developed an operational manual that specifies the roles of the partners, staff members, lead farmers, and contracted farmers as well as their benefits. PRAN personnel take steps to adapt technologies and processes to local conditions. There are no functional linkages between any research/extension group and PRAN at any level. PRAN simply adopts new technologies and seeks subject matter–specific training as needed.

Contracted farmers are organized into farmers' groups, with the average group size varying from 300 to 3,000, depending on location. All groups have a center head or lead farmer. The center head or lead farmer manages and organizes his/her group under the supervision of PRAN.

Problems related to technology, production, marketing, and input supply are discussed and analyzed jointly by the lead farmer, subcenter head farmers, and block supervisors. If problems cannot be resolved by this group, then PRAN management will assist.

Lead farmers receive a commission on their group's produce. Extension agents are incentivized to assist their clients in improving the quality of their crops. Similarly, subject matter specialists and program managers are evaluated and promoted annually. Lead farmers are accountable to the extension agents, extension agents to the specialists, specialists to the program managers, and program managers to the general managers. If the quality of the crops fails to meet the agreed-upon standards, the lead farmer must take the products back with warnings for the next harvest. In the same way, extension agents, specialists, and program managers have similar pecuniary incentives.

Program managers monitor the programs implemented in the field through regular field visits. Farmers also participate in monitoring, from the field to the processing center. It helps them better understand the need for QC in the field and how it affects the quality of products further down the value chain. There is an easy flow of communication from the field to the company and vice versa. Lead farmers are notified about issues by the farmers, and this information is then passed on to the extension workers by the lead farmers, to subject matter specialists by the extension workers, and to program managers by the subject matter specialists, as needed. Farmers receive information from PRAN management through the reverse channel. It should be noted that every farmer has the mobile number of PRAN senior management, including the CEO and the deputy managing director.

Cost sharing and cost recovery

The costs of the extension program are financed by PRAN. The costs of production are shared by the company as necessary, via loans to the farmers. After harvest, the farmer pays back the loan by deducting the value of the loan from the payment for his/her crops. With the program, a farmer can improve his/her production techniques and inputs, especially in the following ways:

1. purchase the best-quality seeds at low prices;
2. receive free training on farm management practices;
3. access financial support during the production period; and
4. obtain irrigation equipment and other required inputs at low prices.

Farmers also benefit from the elimination of middlemen. For these reasons, a farmer contracted with PRAN has more job security than a farmer who only has access to traditional extension.

Program coverage and evaluation

As of 2013, there were 78,000 farmers participating in the program in Natore, Rajshahi, Chapai Nawabgonj, Dinajpur, Gaibandha, Lalmonirhat, Khagrachari, Panchagarh, Munshigonj, Shariatpur, Jessore, Satkhira, and Naogaon districts of Bangladesh. The key impact of the program observed at the field level is increased yield through use of private extension programs. The main constraint to scaling up is convincing farmers of the benefits of the program.

Other agricultural activities such as forestry, fishing, plantation, poultry, or cattle rearing could adopt a similar extension program. External impact evaluations of the program have been conducted by DANIDA, ADB, and Bangladesh Bank (Bangladesh's central bank). PRAN intends to scale up the program in other areas of the country. Thus far, the extension program is considered to be successful, so there is no intention to restructure the program in the near future. From the producers' viewpoint, the crop-specific advice and training can be easily understood and adopted. Every effort has been made to achieve a high level of productivity to enhance the cost-effectiveness of the extension program. The feedback loops from farmer to management and from management to farmer have facilitated continuous learning and refinement of the program.

The private extension system developed and operated by PRAN has been quite successful in improving farmers' knowledge of production and increasing their yield. The extension program covers the costs of accessing information otherwise paid for by the farmers themselves. The program also enables farmers to avoid selling to middlemen who may not offer fair prices.

Partner agencies

Bangladesh's Department of Agricultural Extension rarely communicates with PRAN about field-level activities. Local governments and organizations such as union parishads sometimes interact with farmers through their involvement in the input sales. Research institutions (such as BARI, BINA, and Mango Research Center) have collaborated with PRAN on extension activities in newer areas. Buyer Crop Company, Syngenta, CDCS, and Oxfam have worked with PRAN to supply inputs or to help arrange a supply of inputs.

ANALYSIS OF PRIVATE EXTENSION APPROACH

The PRAN contract farming program has been in operation since 2001. At the beginning, the program had many difficulties, but it has since worked through all the earlier problems. However, the country's sociopolitical unrest presents challenges to sustaining the current extension efforts with the farmers. To the best of observational knowledge, farmers have obviously benefited from the current extension services on several dimensions. As a result, they have shown no intention to get rid of the system, even out of curiosity. The geographical areas covered by PRAN have no competition with any other company; rather, farmers previously engaged with other systems, including the public extension system, have move toward this company.

The model developed by PRAN could be replicated by other companies, both in Bangladesh and abroad. The following sections describe the relevance, effectiveness, sustainability, efficiency, equity, and impact of the program.

PRAN's contract farming programs are relevant because the extension services provided to participating farmers are targeted specifically to the crops they are growing. These programs are effective as they promote a context-specific solution to farmers in a timely and convenient manner. Participating farmers receive sufficient resources such as technical assistance, crop protection chemical products, and financial assistance to improve their yield potential. PRAN's contract farmers are also provided with necessary information and inputs at each stage of the production cycle. The program is considered efficient as it uses its human and financial resources to maximize the benefits to contract farmers through services provided. PRAN's program is considered impactful as contracted farmers have reported significantly increased profits. Higher incomes provide farmers with the opportunity to improve their living conditions, and to access better health care, social services, and higher education. PRAN's extension strategies are considered to be sustainable, as participating farmers have been able to cite tangible benefits such as higher quality of products due to PRAN's quality parameters, less disease, less harvest loss, and lower production costs due to the low price of the quality seed supplied by the company. Farmers have also reported reduced uncertainty of both inputs and outputs as well as yield, namely from reduced disease attacks and price fluctuations. PRAN's programs promote equity, as farmers can participate irrespective of farm size or educational level. PRAN's farmers have equal access to inputs at low prices. A committee comprised of members from PRAN and farmers' representatives decides the prices of outputs, which depend on local market prices. All farmers have access to market information and can make decisions (buying input materials and selling outputs) based on their own cost–benefit analysis.

In terms of the extension workers employed by PRAN, the program is considered to be very relevant. The PRAN contract farming program provides college and university graduates (in agricultural fields) who often have little or no work experience with job opportunities and field experience. The program provides field-level extension workers with transferable skills, such as interpersonal skills (for better collaboration with farmers) as well as information on recent research on recommended farm management/field practices. These skills match those demanded by the agriculture and agribusiness sectors, and therefore PRAN is also effective. PRAN is considered efficient in the eyes of the extension workers it employs as the program has been able to increase the salaries of its extension workers; its business performance has been improving continually since its inception. Extension workers report that the program is sustainable as employees receive regular training, which increases their capacity and updates their knowledge base based on current research. The programs promote equity as all extension workers receive a share of the company's profit, which is determined at the end of the season. In each season, the company gives one extension worker a devotion award for best performance, which helps them become field experts and/ or researchers. The PRAN contract farming program has the intended impact as it contributes quality extension workers to Bangladesh's agricultural sector.

From the viewpoint of management, PRAN is relevant as its programs provide comprehensive services to farmers for crop production to secure a steady supply of inputs for processing. PRAN programs are also considered to be effective as the company has seen improved business performance through the value addition of its extension agents. PRAN has secured a stable and predictable supply of high-quality inputs for agroprocessing, which strengthens their bargaining power against importers of other brands of processed food items. PRAN has developed a profitable line of high-quality agroproducts, which, in turn, helps them further invest in large-scale production for greater economies of scale. In terms of equity, PRAN, as the largest and most important competitor in the agro-processing arena, may be model for others working in parallel. Sharing benefits with its stakeholders, PRAN becomes a reliable partner. The company considers its extension programs to be sustainable as these efforts have developed partnerships with farmers, extension workers, and other stakeholders, and in so doing, has fostered loyalty which may contribute to long-term success. The company also strives to base its services and supplied inputs to match current market demand in order to remain sustainable and relevant. Strong business performance allows PRAN to improve infrastructure and offer better social services to local communities as part of its corporate social responsibility.

COMPARISON BETWEEN PUBLIC AND PRIVATE EXTENSION SERVICES IN BANGLADESH

The private extension approach has opened a new dimension for development in Bangladesh. Public–private partnerships are playing a vital role not only in agriculture, but also in community development, education, industry, communication, and health. Table 11.3 demonstrates that public extension does not follow the value chain approach, engage in contract farming, use ICT, provide inputs, offer credit, nor engage in post-harvest processing. On the other hand, private extension programs include all the activities required by the farmers' production system.

Table 11.3 Comparison between public and private extension approaches in Bangladesh

Extension ownership		Approach			Type of information			Communication method used in extension activities						Other services provided		
		Top-down	Value chain extension system	Research-cum-extension system	Farming system	Environmental conservation	Food safety	Short training course	Extension workshops	Radio and TV broadcasting	Contract farming	ICT application	Technical consulting	Provision of inputs and Marketing of outputs	Credit	Post-harvest processing
Public extension		×	–	×	×	×	×	×	×	×	–	–	×	–	–	–
Private extension	PRAN AL	–	×	–	×	×	×	×	–	×	×	×	×	×	×	×
	Lal Teer Seed Company	–	×	×	–	×	–	×	×	×	×	×	–	×	×	×
	BRAC	–	×	×		×	×	×	×	–	×	×	×	×	×	–
	Syngenta	–	×	–	×	×	×	×	×	–	×	×	×	×	–	×

Source: Author's compilation.

UNIQUENESS OF THE MODEL

PRAN is the largest private-sector contract farming company that is both a manufacturer and a distributor of processed agro-products in Bangladesh and abroad. It has a well-defined field-level network to facilitate training for farmers as well as distribute inputs, credit, and information. PRAN's well-organized input distribution and crop collection system of products have made the contract farming system unique.

The agricultural hub and lead farmer as the focal points of the extension program help the contracted farmers get all their production advice and solutions efficiently. In addition, PRAN has extensive distribution networks, flexible sales policies, and a strong base of loyal customers—in part because of offering direct technical assistance to farmers. Its volume of sales has kept increasing over the last few years. The company's exports have been increasing, not only in terms of geographical area, but also in value. In 2010, the growth rate was approximately 33%, and in 2011 it was 47%. The company is a large distributor of crop protection products throughout Bangladesh. Furthermore, the company's involvement in securing seed supply, paddy collection, processing and storage, and transportation helps to provide complete crop solutions along the value chain for farmers.

LESSONS FOR REPLICATION AND SCALING UP

The PRAN extension program has been in operation since 2001, reaching 78,000 farmers in 13 districts in Bangladesh by 2013. The company plans to double the coverage of its current program. The crops it currently buys are tomato, mung beans, peanuts, mango, rice, chili, coriander, turmeric, tamarind, and olives. It appears that its future plans are feasible given that its contracted farmers have not previously been serviced with a functional extension system.

One challenge that may affect the sustainability of the program is the farmers' side-selling of outputs, despite having a contract with the company. Under the current contract, farmers who want to leave the program must pay for all inputs already received. In addition, farmers who break the contract due to side-selling or other illegal activities are blacklisted from future involvement.

Despite this challenge, the PRAN contract grower program can be replicated by other companies because it is simple and profitable. The corporate mission of the company states that "poverty and hunger are curses" and its aim is to "generate employment and earn dignity for our compatriots through profitable enterprises." The PRAN contract grower program aims to increase self-reliance among farmers.

Moreover, the individual capacities of the contract farmers are strengthened as they engage in more commercially oriented farming. The extension program of the company uses a cost-effective approach to strengthen the individual capacity of the farmers using their inherent potentials. There is a provision to include 10% female members of rural communities as they need significant support in order to benefit from contract farming.

PRAN organizes agricultural hubs that are led by lead farmers with the cooperation of the contracted farmers. It provides quality production inputs directly to its farmers at a low price, provides comprehensive support, decides upon a market price in a transparent way, and collect outputs. These arrangements have helped to establish trust between the farmers and PRAN. This model can be followed by other companies in the country and abroad, and can be applied to other crops as well.

CONCLUSIONS

The public extension system in Bangladesh is filled with weaknesses that make it an ineffective support system for farmers. One of the main areas where farmers do not receive support is at harvest time, when markets are so oversupplied that farmers cannot sell all of their produce. Public extension workers receive a government paycheck, but there is little accountability to the farmers they are supposed to serve. There is a tendency to provide more services to the richer, large-scale farmers, who are often better connected and may have some influence over the extensionists' job security. However, poorer, smaller-scale farmers who are in more need of extension services tend to be neglected.

Because of the unreliability of the public extension system in Bangladesh, PRAN's contract farmer approach is appealing to many farmers. With PRAN's extension services, farmers can interact with a local lead farmer and receive support on a variety of production-related activities. The system facilitates the timely distribution of seeds and other inputs at low cost and provides farm management training. It vows to help farmers find solutions to any issues related to production. PRAN also provides credit to farmers who need assistance purchasing inputs and equipment. This system strengthens the relationships between farmers and PRAN because both parties see clear benefits. These relationships can be further strengthened and the system expanded to other areas of the country to facilitate poverty reduction. The government can play a role in identifying gaps in the extension services provided by PRAN, and can aid in filling these gaps with its own programs and extension services.

REFERENCES

Anon, 2013. WTO agreement on agriculture: potentials of agro-processing products of Bangladesh. Economic Policy Paper.
BAPA (Bangladesh Agro-Processors' Association), 2013. Available at: <http://bapabd.org/> (accessed 21.09.13).
PRAN, 2013. <http://www.pranfoods.net/achievment.php> (accessed 22.09.2013).
Quddus, M.A., Mia, M.M.U., 2010. Agricultural Research Priority: Vision—2030 and beyond. Final Report. Bangladesh Agricultural Research Council.
World Bank, 2012. Rural population (% of total population) in Bangladesh. Available at: <http://www.tradingeconomics.com/bangladesh/rural-population-percent-of-total-population-wb-data.html>.

CHAPTER 12

Extension and Advisory Services for Organic Basmati Rice Production in Jammu and Kashmir, India: A Case Study of Sarveshwar Organic Foods Ltd.

Rakesh Nanda[1], Rakesh Sharma[2], Vinod Gupta[2] and Gaytri Tandon[3]
[1]Division of Agricultural Extension Education, Sher-e-Kashmir University of Agricultural Sciences and Technology of Jammu, Jammu, India
[2]Krishi Vigyan Kendra, Sher-e-Kashmir University of Agricultural Sciences and Technology of Jammu, Jammu, India
[3]Sarveshwar Organic Foods, Jammu, India

Contents

Knowledge Driven Development.
DOI: http://dx.doi.org/10.1016/B978-0-12-802231-3.00012-7

219

INTRODUCTION

Food consumption patterns have been shifting globally, with an emphasis on more processed and specialized commodities such as organic products. In developing countries, production is shifting from food grains to high-value crops and animal products. However, in developed countries, consumer demand is shifting from animal and fish flesh to fresh fruits and vegetables, and within the cereals subgroup from wheat to rice and maize (Goel, 2010). Changing food consumption patterns have provided opportunities for higher-value crops and expanded market size, both domestically and internationally. Export of high-quality rice from India is one such opportunity, which in turn demands high level of engagement of rice farmers with the private sector and depends on an increased knowledge base derived from outside sources, allowing private extension to operate and benefit the farmers. In this chapter we document the extension and advisory services provided to organic rice farmers by the Sarveshwar Organic Foods Company in the Jammu region of India.

BASMATI RICE AND ITS EXPORT FROM INDIA

Basmati rice (*Oryza sativa*), known as the king of rices, is priced for its characteristic long grain and subtle aroma and taste. Basmati rice is one of the major agricultural commodities being exported every year to earn foreign exchange (Kumar and Singh, 2011). It is cultivated primarily in India and Pakistan. India is the largest producer of basmati rice, accounting for about 70% of world production. Domestic production was approximately 3.35 million (mn) tons during 2008–2009, 6.4 mn tons during 2009–2010, 6.5 mn tons in 2010–2011, and 7.5 mn tons in 2011–2012. In India, basmati has been cultivated for thousands of years on the Himalayan foothills in the states of Jammu and Kashmir (J&K) in Jammu region, and in Himachal Pradesh, Punjab, Haryana, Delhi, Uttaranchal, and western Uttar Pradesh (APEDA website). This chapter will focus specifically on the production and extension efforts in J&K states using Sarveshwar Organic Foods Ltd. as a case study.

Rice is one of the three most important food crops in the world, forming the staple diet for 2.7 billion people. It occupies 150 mn ha, producing 573 mn tons of rice, with average productivity of 3.83 tons/ha. In India rice accounts for 40% of food grain production. With liberalization of the economy during early 1990s, the government of India undertook several initiatives to enhance its exports, including financial assistance to exporters to improve basmati rice quality, packaging and brand promotion, encouragement of participation in international fairs, and organization of buyer–seller meetings.

In October 2008, the central government pegged its exports of basmati rice with a minimum export price (MEP) of US$900 per ton. Up to this time there had been no fixed MEP for basmati rice and exports were restricted to the ports of Kolkata, Kandla,

Kakinada, Navi Mumbai, Mundra, and Pipava. In addition to national regulation, individual state governments made amendments in the Agricultural Produce Marketing Committee (APMC) acts to increase opportunities for contract farming because of the existence of Land Ceilings Act, which stipulated that agribusiness firms could not own and cultivate land for raw materials. Changes in policy at both the central and state levels have enabled the industry to introduce farm-level changes and build up its competitive edge. This has drawn several corporate groups, MNCs, agri-input agencies, and other organizations to contract farming, though its models have varied between crops, regions, and sponsoring organizations. As a result, the average share of basmati rice exports in total agricultural exports, which stood at 5.8% during the period 2001–2002 to 2007–2008, rapidly increased to 11% during 2008–2009 and further to 12.1% during 2009–2010. The most common brands of basmati rice include Sarveshwar, Lal Qilla, Double Diamond, Daawat, Kohinoor, Doo, and Annapurna. Sarveshwar Organic Foods is one such company and has been aggregating the basmati rice from the farmers for several years. Sarveshwar operates in J&K, Uttar Pradesh, and Himachal Pradesh, covering an area of 5,000 ha of basmati rice cultivation (Table 12.1).

Basmati rice exports from India have moved up in a fluctuating manner in volume and value terms between 1991–1992 and 2009–2010 (Figure 12.1). Implicit export

Table 12.1 Basmati rice cultivating districts of India

India state	Districts
J&K	Jammu and Kashmir
Haryana and Punjab	Karnal, Panipat, Kurukshetra, Kaithal, Amritsar, Fatehgarh, Gurdaspur, Basmati Hoshiarpur, Jalandhar, Patiala, Sangrur, Roopnagar
Himachal Pradesh	Kangra, Solan, Una, Mandi, Sirmour
Rajasthan	Bundi
Uttar Pradesh	Saharanpur, Muzaffarnagar, Pilibhit, Bareilly, Bijnor, Moradabad, Jyotibaphule Nagar, Rampur, Sitapur, Rae Bareli
Uttarakhand	Udham Singh Nagar, Haridwar, Dehradun

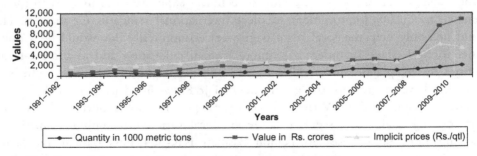

Figure 12.1 Basmati rice exports from India. *Source: www.apeda.gov.in.*

Table 12.2 India's basmati rice exports

Year	Absolute values			Change (%)		
	Quantity million (MT)	Value (billion Rs.)	Implicit prices (Rs./qtl)	Quantity	Value	Implicit prices
2001–2002	0.67	18.43	2,762.48	−21.68	−13.92	8.63
2002–2003	0.71	20.58	2,904.20	6.25	11.71	5.13
2003–2004	0.77	19.93	2,583.38	8.85	−3.18	−11.05
2004–2005	1.16	28.24	2,428.12	50.75	41.69	−6.01
2005–2006	1.17	30.43	2,608.59	0.31	7.76	7.43
2006–2007	1.05	27.93	2,670.68	−10.36	−8.22	2.38
2007–2008	1.18	43.45	3,671.39	12.16	55.56	37.47
2008–2009	1.56	94.77	6,089.03	31.52	118.13	65.85
2009–2010	2.02	108.39	5,376.66	29.52	12.37	−11.70

Source: From Goel (2010).

price, and thereby its value realization, have seen a steep jump in the last couple of years. Table 12.2 illustrates the trends in the value and quantity of basmati rice exports from India.

PRODUCTION OF BASMATI IN JAMMU DIVISION

Basmati rice is a scented variety of superfine rice grown in subtropical areas of the state, namely R.S. Pura, Bishnah, Kashmir, Jammu, Akhnoor, Samba, Hiranagar, and Kathua Tehsils of Jammu division. Cultivation of basmati rice under a diversified cropping system is ideally suitable due to its low water requirement and low susceptibility to insect pests and disease. There is good scope for areal expansion of the basmati crop. It is possible to increase the production of basmati rice in the state and generate more potential for export of the scented varieties of this crop, provided that some central export agency such as APEDA or some other export organization in the public or private sector extends its technical expertise to farmers, with special reference to quality improvement to meet international standards for exports. This would not only help the local basmati growers economically but would be a dollar earner for J&K state. In order to promote basmati cultivation and to protect farmers from the malpractices of middlemen and millers, and to give the maximum benefit to the basmati growers, the government of J&K has lifted its ban on the export of basmati rice as of 2009. Varieties having great potential for export are known for their cooking quality and scented nature are local varieties, including Basmati 370, Puse Sugandha, Sanwal Basmati, Ranbir Basmati, RR-564, Pusa Basmati no. 1, and Basmati 1121.

OVERVIEW OF SARVESHWAR ORGANIC FOODS LTD. AND ITS ROLE IN ORGANIC RICE PRODUCTION

Organic production systems are based on highly specific standards precisely formulated for food production; they aim at achieving an agroecological system that is both socially acceptable and ecologically sustainable. Organic farming uses ecological principles as the basis of crop management and animal husbandry. The Codex Alimentaris Commission, a joint body of the Food and Agriculture Organization (FAO) and the World Health Organization (WHO), defines *organic agriculture* as a holistic food production management system which promotes and enhances health of agroecological systems, including biodiversity and soil biological activity, and emphasizes the use of management practices in preference to the use of off-farm synthetic inputs.

Sarveshwar Organic Foods Ltd. is a part of the Sarveshwar group of companies. The group has been in business for over 58 years, with its headquarters in Jammu, India. Sarveshwar Organic Foods Ltd. was established in 2004 under the Indian company act and is involved in the farming, processing, and export and marketing of organic products. It has led the effort to promote organic production in J&K. It has developed a broad customer base in the European Union countries, the United States, Canada, Australia, New Zealand, and the Middle East, and it looks forward to extending its operations to other developed countries as well. The headquarters of Sarveshwar Organic Foods Ltd. are situated in the city of Jammu district.

Sarveshwar organic project in Jammu is principally a smallholder support project. In 2013 there were over 5,000 farm families growing basmati rice organically in 157 villages of Jammu, Samba, and Kathua districts of Jammu division of J&K on 4,900 ha of land. Under this project, farmers follow organic production and processing practices as recommended by the company. Participating farmers are provided with appropriate training regarding organic concepts and their limitations. A majority of the farmers are smallholders, and many are below the poverty line. The average size of these farms is 2–4 ha; the average area under basmati rice is 1–2 ha per farm. The farms are mainly managed by family members with the additional labor of seasonal workers during transplanting and harvesting periods. Farmers sell a portion of their rice to Sarveshwar Organic Foods Ltd. and retain the remainder for household consumption and as seed for the next season.

DESCRIPTION OF THE EXTENSION APPROACH OF SARVESHWAR ORGANIC FOODS LTD.

Sarveshwar Organic Foods Ltd. conducts several training-cum-awareness generation programs to generate quality assurance among its farmers. Farmers are instructed in the use of compost, decomposed organic matter, and green manure as natural fertilizers and local formulation of approved bio-chemicals for pest and disease control.

The activities of the company began with a baseline survey to collect primary data such as identification of suitable areas. Representatives from the company then attended village-level meetings regarding the program for the purpose of selection of villages, farmers, and areas. Following that they organized farmers into groups and groups into societies. The primary data was collected by conducting participatory rural appraisal of the selected villages. The secondary data was collected from revenue records of selected villages, areas, and farmers selected to find cropping patterns over the previous 3 years; the inputs being used by the farmers; basic facilities available to the farmers with respect to organic farming; and interventions required (construction of compost pits, infrastructure for biological/botanical preparations, change in cropping patterns, etc.). Village maps are prepared to depict organic areas of individual farmers and high-risk areas due to aerial drifting from conventional farms, as well as buffer zones. Individual farm maps are also prepared to depict crop rotation and buffer zones. These materials are helpful to extension efforts, making them more tailored and relevant to the farmers' unique needs.

Activities for promotion of organic rice

The main objective of training is to inform and train project staff and organic growers on the relevant aspects of organic farming and, in particular, to make them aware of the content and practical implications of the regulations for organic agriculture.

Training of internal control system (ICS) personnel

In order to conduct project activities, a team is established and trained, including a project officer and project executives. All field officers are trained once a year (before the beginning of the new control season, usually in April). This updates their knowledge base and helps their advice remain relevant to current research. All officers are trained for organic farming, practices to be followed in the organic field, input management, documentation, and organic certification.

Training of farmers

To effectively train the farmers covered under the project, customized farmer training sessions are conducted. During these sessions, topics are covered in detail. Training programs are planned in advance and offered to each and every person of a village. Farmers are given thorough training on the following topics:
1. Organic crop production techniques
2. On-farm input production and judicious use of natural resources
3. Resource conservation and optimization
4. Preparation of on-farm organic manure
5. Vermin-composting, cocoon and culture multiplication, and proper utilization of vermi-compost

6. Maintaining records for all operations on the farm
7. Pest and disease management by organic methods
8. Post-harvest management
9. Packaging and labeling

Training farmers on the production of rice

Harvesting season for rice begins around November 15 each year. Initial field visits are made and farmers are trained in changes in the methods of harvesting, including the following techniques:

- Rice crop should be harvested when the grains become hard and contain about 20–22% moisture.
- Harvesting before maturity means a low milling recovery and also a higher proportion of immature seeds, high percentage of broken rice, poor grain quality, and more chance of disease attack during storage of grain.
- Delay in harvesting results in grain shattering and cracking in the husk, and exposes the crop to attack by insects, rodents, birds, and pests.
- Avoid harvesting during wet weather conditions.
- Drain the water from rice fields about a week or 10 days before the expected harvest, which helps in employing mechanical harvesters.
- Keep the harvested rice separately for each variety, to ensure true to type variety.
- Avoid direct sun drying, which leads to an increase in breakage of the grains during milling.
- Pack the rice in sound B-twill jute bags totally free from any contamination.

Jute bags are provided to the farmers for storing rice. A separate counter is opened at each cluster for purchase of rice. Two permanent staff are assigned to be available to assist farmers throughout the week. Farmers under contract are given a premium price for organic rice.

Activities conducted by ICS personnel

Many of the extension activities that are helpful for the farmers relate to the production and processing of organic rice. The following activities are undertaken by ICS of the Sarveshwar Organic Foods Ltd.

Awareness generation and cluster formation

After the appointment and orientation of project staff, they are deputed in the field to create awareness by visiting the villages, holding group meetings with farmers, and discussing the concept of organic farming and its benefits, average holdings, and the need to form clusters for organic certification under the small farmer group scheme. Project staff had meetings with field staff of the agriculture department. They visited various villages and met with registered farmers.

Preparation and display of publicity material

Publicity materials in the form of panels and banners were developed and displayed at the entrance of each village. This helped to promote the program and its credibility to local farmers. Further, it introduced Sarveshwar Organic Foods to the farmers, thereby promoting better trust between the farmers and the company.

Cluster formation

In order to form small farmers groups, i.e., clusters, for organic certification, landholding is the main component. The local land unit prevalent in Jammu division is the kanal, which is a colloquial unit; 20 kanals equals 1 ha. After registration of farmers, a certification agency is finalized, called the Control Union. This agency is accredited for inspection and certification of organic products by APEDA under NPOP guidelines of the Department of Commerce, Ministry of Commerce and Industry, Government of India.

Preparation of field diary and its maintenance

In order to record daily activities conducted in the field, diaries are prepared and distributed to every farmer as part of the certification kit. The diary contains the details with respect to the farms, describing the details on other farms as well, operations/ activities performed on the farm, details on nutrient management and plant protection measures followed, cattle, harvesting, and tentative yield records. To complete the records, the project staff and ICS persons visited every farm and completed the details. In order to check the activities performed and ensure proper documentation of the farm diary, checklist, etc., internal inspection was also to be done.

Internal inspection

Internal inspections of grower groups are carried out by internal inspectors twice in a year. Internal inspectors inspect all the production practices covered by the ICS, including production, plant protection, and intercultivation operations.

External inspection

External inspection is carried out by the certification agency once a year. A second external inspection is carried out and, a C-2 certificate is issued. During the external inspection by the Control Union, the effectiveness of the ICS is evaluated. The external inspector re-inspects a certain number of farmers under the certification agency's policy. The percentage of external control will be determined by the certifier on the basis of a risk assessment. Also, the inspector may undertake witness audits; that is, he accompanies the field officers to evaluate the effectiveness of their inspections. The external inspector compares his observations with the control documents and evaluates the ICS. The ICS will be certified only after fulfilling the requirements (entrance application, signed contract, updated plot maps, and all internal inspection reports) of respective standards.

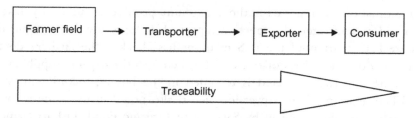

Figure 12.2 Operational model of Sarveshwar Organic Foods Ltd. *Source: Author's compilation.*

In case corrective measures are suggested by the external certification agency, they must be communicated to the concerned field officers to implement.

Storage of organic produce and maintenance of traceability

All organic produce procured from farmers is stored at a separate place, and this designated area is declared an organic zone. From field to storage, organic rice is carried in different vehicles, and a separate stock register is maintained for entry. The traceability record from farm to fork is maintained as shown in Figure 12.2.

UNIQUENESS OF THE MODEL

The company purchases rice from the farmers on the premium rates on a buyback basis. Buying is organized in local centers (*mandis*) by the local *arti* (middlemen), pre-recognized by Sarveshwar Organic Foods Ltd., in the presence of the company's own buying officers. The buying process starts in front of the buying officers of the company, who are well trained and are full-time staff of company. Sarveshwar Organic Foods Ltd. maintains at least one such officer at each buying station. Weights are taken and a tag is fixed to each bag with details of the contents (farmer code, weight, status, variety). The buying officers are in charge of ensuring correct purchase of organic produce from the farmers. This way the company maintains an accurate organic product flow control, as even a single mistake at the purchase level can have a huge impact on the organic project. For raw organic rice, a registered farmer is identified as an organic farmer or a farmer in the process of conversion. Sarveshwar Organic Foods Ltd. provides separately marked jute bags to each grower according to predicted yield. The rice is then bagged and brought to the purchasing center (the factory). In each center, in front of the buying staff, weight is measured and gunny sacks are issued. Rice from in-conversion farmers is bought and marketed under a separate marketing plan, and is handled with the same level of care as followed for a fully organic rice. Specific care is taken to avoid contamination and commingling of organic and conventional rice. The purchasing officer confirms that the delivered quantity of organic rice is plausible by comparing the actual delivery with the yield estimate.

Yield estimation is done with the maximum possible accuracy by taking past yield data and present internal inspection reports. If there is any doubt about the rice, the bags are kept apart until the ICS manager has checked the farmer's documents. After confirmation, those bags of rice are absorbed into the organic supply chain. The quantity of organic rice delivered is recorded in the buying record with statement "organic." The quantity of in-conversion rice is also noted in the same record book. A cash voucher (receipt) is given by Sarveshwar Organic Foods Ltd. to farmers with details of the farmer's name, farmer's father's name, village, quantity supplied, and total amount paid. The bags of each organic farmer are labeled with their certification status (i.e., I, II, Organic). After the buying procedure, the collected products are transported in hired trucks from the purchase center to the central warehouse of the Sarveshwar Organic Foods Ltd. These bags are stored in a specified warehouse and are kept separately from conventional rice bags in a controlled environment. From there, rice bags are taken to the processing plant for milling. Before each milling, it is verified that the machines are empty and clean. The company runs its own processing unit and specifically assigns a well-trained processing manager to supervise organic handling procedures. He is fully responsible for organic processing according to the internal rules. The manager personally supervises the whole process to avoid mixing. After processing, the rice is packed in 25 kg bags. Milling is done only on order. After packing, goods are loaded in containers and dispatched from the factory. From the moment of purchase from the farmer until export, the rice is owned by Sarveshwar Organic Foods Ltd.

Registration of new farmers

If a farmer wants to join the organic program, he needs to apply for admission before the sowing season starts. He needs to declare all of his plots (rice and others) to ICS and provide the date of last application of agro-chemicals. Farmers who already participate in the organic program but have new fields also need to declare the last use of non-allowed inputs, as a period of 3 years since the last application is required for the land to be allowed under organic production. Each new farmer has to complete a training program in organic production during the first year of registration. The farmer has to declare openly all cultivation measures, purchase of inputs, and treatments applied on these fields. The farmer must take great care to prevent drift of chemicals from his own conventional areas or neighboring fields. The farmer is not allowed to use any off-farm inputs on his organic plots (e.g., fertilizers, insecticides, fungicides, herbicides) except those for which he has been explicitly been granted permission by the internal inspector. Naturally grown (botanical) fertilizers/pesticides can be used, but these also need to be declared to ICS before use. Fertilization may only be done with dung, compost, and FYM. For plant protection of rice and other crops cultivated in the organic farming unit, products based on organic neem (*Azadirachta indica*) are authorized. Only those commercial products may be used that have been authorized by the ICS/certifier. The farmer must store the inputs used in the conventional farming unit in such a manner that no contamination by organic inputs or crops can take place. He must allow access to internal and external inspectors to verify the storage of inputs during inspection. In order to avoid any contamination of the organic fields, Sarveshwar-registered farmers are asked not to use spraying equipment hired from conventional neighbors and are also advised not to lend their equipment to conventional neighbors. The farmer is obliged to ensure soil fertility by using appropriate cultivation measures and to minimize erosion. Each new farmer in an organic farmers group must send a completed ICS application form to Sarveshwar Organic Foods Ltd. field office. If, after an initial document review, it is found that the farm fulfills the group certification norms set by NPOP/EU regulations, the farmer will receive a field inspection by an ICS inspector. Only small farmers can be members of the group covered by group certification. Larger farms (i.e., farms bearing an external certification cost that is <2% of their turnover) can also belong to the group, but these must be inspected annually by the external inspection body. Processors and exporters can be part of the structure of the group, but must also be inspected annually by the external inspection body. The farmers in the group must apply similar production systems, and the farms should be in geographic proximity. A group may be organized on its own, i.e., as a cooperative or as a structured group of producers affiliated with a processor or exporter. The group must be established formally, based on written agreements with its members. It must have a central management, established decision-making procedures, and legal capacity. When intended for export, the marketing of the products must be carried out as a group.

Advisory services rendered by the company to organic rice farmers

The requirements and obligations of organic production and contracting with the company are explained to the farmers up front. An ICS inspector assists the farmer in maintaining the farmer's diary. The inspector will draw a simple sketch of all the plots belonging to the farmer, and a rough indication of present crops and potential risks of drift. The farmer signs the contract with the company's Organic Farmers Group. The ICS inspector fills out the ICS inspection form after a full farm inspection. Once the documentation is complete, the information is processed by the ICS office. The inspection form is then screened by the organic manager and the conversion status is determined according to the internal organic standards. Once accepted, the farmer receives a farmer code number, which appears on all documents relating to this farmer. The farmer's information is added to the list of registered farmers. The organic farmers group is finalized, and the inspection reports are handed over to the organic approval committee, which scrutinizes the results. It checks the fulfillment of the previous year's conditions and the new conditions proposed by the internal inspector. It decides on approval or sanction of each producer and determines the conditions and duration of any sanctions.

Conditions and sanctions are registered in the farm inspection report. The results of the meeting are summarized in the list of approved and sanctioned farmers.

Additionally, there is a short protocol of each meeting of the approval committee. If a grower in the organic farmers group violates the internal standards, appropriate sanctions and corrective measures are determined according to the list of nonconformities. In the case of severe violations, the internal inspector or whoever detects the incident must fill out a violation report. Sanctioned producers may not sell their produce to the Sarveshwar Organic Foods Ltd. for the whole period of sanction. The reason for and duration of the sanction is noted on the list of sanctioned producers and the purchase officer is informed accordingly.

The documents of each farmer of the organic farmers group are kept in the individual farm files, which are stored in the office of internal control (organic project sites). The farm files for each farmer contain the documents that have been produced during the registration procedures plus the farm inspection checklists. The data for all farmers and the results of the internal control are summarized in the farmer list and the list of sanctioned farmers.

ANALYSIS OF PRIVATE EXTENSION

Interviews with farmers growing organic basmati rice under the Sarveshwar scheme have shown increased awareness of organic production techniques. Although the chemical fertilizer use in basmati production was lower in the region compared to the national average, the acquired knowledge about rice production under organic conditions has helped farmers to understand the sustainability issues related to rice production.

To be successful, it is critical to moblize farmers to organize themselves to produce basmati rice that will meet the standards for organic products, which can then be exported. This has increased the relevance of the messages from the private company. The provision of training to farmers throughout the crop season has increased the effectiveness of the extension system as the extension workers focused on specific information needed for the different stages of crop production. As the selection of farmers focused on smallholders in the region, organic production of basmati rice with the help of the private company has helped to increase the equity among farmers in terms of reaching out to all types of farmers in the region, some of whom would not have benefited through the regular public extension system.

Since the farmers were educated in the inspection aspects of organic production, compliance with the regulations for export of organic rice increased. Meeting the high standards of organic production was initially a challenge for the farmers. However, the role of Sarveshwar Foods in enabling organic production of rice by registering the farmers and ensuring that the farmers followed the steps needed for certification increased the compliance rate, thereby enabling the export of their rice by the company. The company paid a higher premium for this rice, which helped to increase the income of the farmers.

Organic rice production should be sustainable as long as the farmers are organized by the company and regularly trained in the organic production of rice. The training and monitoring of the farmers need to continue for sustainability of the production and extension approach.

LESSONS LEARNED

Several factors that helped the company's approach to contracting and extension provision are listed below:

- Extension services provided by the company are a key factor that determined the success of the contractual arrangement as the compliance of the farmers to the organic standards depended on the training provided by the company.
- A key lesson from the experience of the company is that if the farmers could be organized toward specific goals in production and provided adequate support in terms of assured markets, the chances of success could be higher.
- The higher premium paid for the rice by the company attracted the farmers to take up organic rice. This interest should be there as long as farmers get higher rices.
- Organic rice production is highly knowledge intensive, and thus the training of the farmers through extension provided by the company is key to the success of the program.
- The compliance of farmers with organic standards increased due to the extension services provided in the compliance procedures.
- Farmers as well as the company benefited through the contractual arrangement, and this is a win–win situation that is being replicated and scaled up by the company.
- Other players can also achieve success if they follow a similar approach. Encouragement of the farmers by public extension workers to join the organic production of rice has also helped to gain the confidence of the company. This public–private cooperation has been helpful in allowing other companies to enter contractual arrangements with farmers.

CONCLUSIONS

In this chapter, we presented a case study of the extension services provided by Sarveshwar Organic Foods Ltd. for the promotion of organic basmati rice in India in order to understand how a private-sector organization provides extension to its farmers. The areas highlighted in this chapter are those where the rice–wheat cropping system has been widely practiced. The higher levels of productivity and stability of this system provided employment and income to rural masses and also ensured food security in the region. However, this system is now showing signs of fatigue. Yield stagnation and declining factors of productivity are eroding the profit margin of the farmers. Moreover, there has been enormous damage to natural resources. Declining soil fertility due to excessive use of chemical fertilizers and an increasing problem of insect pests and diseases in the rice crop make the situation complex and very serious. These problems highlight the need to convert production to a system which is more environmentally-friendly and sustainable, and that creates income opportunities for farmers. Based on the findings of existing practices of paddy in the region, new management practices were devised and technology inputs were given to the farmers. These standards provided farmers a scientific solution for crop management and creation of safe and quality brands for better market acceptance and opportunities under a free trade environment.

In order to impart training to the farmers covered under the project, customized farmer training is conducted. These training programs were planned in advance for farmers in line for organic production: bio-pesticide application according to legal requirements, resource conservation and optimization, preparation of on-farm organic manures such as CPP, vermi-composting, cocoon and culture multiplication, proper utilization of vermi-compost, and record management for all farm operations. To convince farmers to cultivate organic basmati, the company fosters a strong bond with the basmati growers. The success of the company is due to the buyback facility provided to the farmers, and at the premium rates paid—the company purchases the crop from farmers' fields at a rate that is 10–20% higher than the market rate. This intervention paves the way for many other private players in helping farmers to get remunerative prices on a sustainable basis without affecting the ecology.

REFERENCES

Goel, V., 2010. Recent shifts in global food consumption patterns and future scenario. 20th Annual World Forum and Symposium on Navigating the Global Food System in a New Era held on 19th–22nd, 2010. Boston, MA.

Kumar, J., Singh, R.P. (eds), 2011. Proceedings of the 25th Training on Quality Management and Plant Protection Practices for Enhanced Competitiveness in Agricultural Export. G.B. Pant University of Agriculture and Technology, Pantnagar, Uttarakhand, India.

CHAPTER 13

Private Extension Provision in Vietnam: A Case Study of An Giang Plant Protection Joint Stock Company

Pham Hoang Ngan[1], Tran Tri Dung[2,3] and Suresh Chandra Babu[4]

[1]Vietnam Inclusive Innovation Project, Hanoi, Vietnam
[2]DHVP Research and Consultancy, Hanoi, Vietnam
[3]Centre for Creativity and Innovation, Boise State University, Boise, ID, USA
[4]International Food Policy Research Institute (IFPRI), Washington, DC, USA

Contents

Knowledge Driven Development.
DOI: http://dx.doi.org/10.1016/B978-0-12-802231-3.00013-9

INTRODUCTION

This chapter outlines an example of tailored technical assistance provided by a private company in Vietnam to increase the knowledge and livelihoods of small-scale rice farmers. The production, domestic consumption, and export of rice play a vital role in Vietnam's agricultural development strategy. Improving the rice productivity of smallholder farmers and connecting them to domestic and export markets remains one of the government's major priorities for the agricultural sector. Agricultural extension, as designed and implemented by the Ministry of Agriculture and Rural Development (MARD) and the provincial governments, plays an important role in the ongoing transformation of the rice sector. This program is centrally run, and its top–down approach often leads to bureaucratic inefficiencies. In addition, public extension programs in Vietnam do not have the human or financial resources to meet the diverse needs of smallholders. Due to the high demand for knowledge and information services from farmers and the government's limited resources, the private sector and NGOs have become increasingly active in providing farmers with extension and advisory services. This case study documents the operations and challenges of the An Giang Plant Protection Joint Stock Company (AGPPS) in providing extension services to its stakeholder farmers.

PRIVATE EXTENSION BY AGPPS

AGPPS has a variety of different projects and programs that aim to meet the needs of a diverse set of rice farmers. The next section provides an overview of one of the company's many agricultural extension programs, farmers' friends (FFs).

FFs program

AGPPS, which was established in 1993, is a provider of seedlings, fertilizers, and pesticides and is a conduit for the transfer of agricultural technologies and cultivation techniques in the rice industry. Beginning in 2007, in order to address major agricultural pests and diseases threatening Vietnamese rice producers, AGPPS selected well-informed farmers to act as FFs. AGPPS was the first private organization to provide services to smallholder farmers directly in the field through its FF program. During the early stages of the program, the program assisted farmers in the proper use of pesticides to protect paddy fields. Initially, there were just 12 farmers enrolled in the program. The team worked with farmers in three demonstration fields (termed illustration models) and in 146 family farm plots where seminars were given (termed illustration points). Each FF worked with about 20 farmers.

In 2011, AGPPS integrated its FF program with its investment project in large-scale rice cultivation (LSRC). The company also launched its food industry development project, aiming to facilitate the sale of rice, by branding Vietnamese rice through an integrated model of producing and selling. The collaboration between these programs expanded the knowledge base of both the extension providers and the farmers participating in the program.

At present, there are about 424 FFs working at 74 illustration models, 1,993 illustration points, and 42 technical consultation points. Technical consultation points are centers staffed by FFs where farmers can stop in for advice or consultations. Table 13.1 describes how technical consultation points, illustration models, and illustration points are used to engage with farmers. It is hypothesized that technical assistance provided by FFs has dramatically improved since the LSRC project began, both in quantity and quality.

LSRC in An Giang province

In 2011, AGPPS started large-scale cultivation on more than 1,100 ha of paddy fields, which accounts for a third of the rice cultivation area in the province of An Giang. AGPPS is the pioneer of LSRC in Vietnam. In 2011, the province applied its cultivation model to 3,867 ha—the largest area yet—which involved the participation of more than 3,260 farming families.

Vietnamese Good Agricultural Practices (VietGAP) procedures and standards were employed in order to improve quality assurance and increase market access for participating farmers. VietGAP was introduced by MARD on January 28, 2008. It is an accreditation that is issued for agricultural products that meet four production criteria: good production techniques, food safety (no bacterial contamination, or chemical or physical pollution during harvesting), good working environment (no labor abuse), and a traceable chain of production.

Farmers receive support through the LSRC program in five ways. AGPPS sells necessary inputs such as seedlings, fertilizer, and pesticides to farmers at cost. After harvest, AGPPS sells the rice on behalf of the farmers, paying them their revenue minus the cost of the provided inputs. Second, AGPPS sends technical staff to work directly with farmers in the fields, providing them with information about cultivation, harvest,

Table 13.1 Types of technical assistance provided to farmers

	Technical consultation points	Illustration models (demonstration farms)	Illustration points (family plots where demonstrations are held)
Description	A technical consultation point is a permanent consultation center staffed by an FF. It doubles as a hostel for the FF. In the future, the point will also host a radio station to share information and connect with farmers.	There are two types of models: – Large-scale production field. – Production cluster, where tens of farmer families work together on 10–100 ha. At the model, FFs work with farmers in the fields and organize on-field seminars to disseminate information about natural enemies and disasters as well as to educate farmers on crop management.	An illustration point is a client's farm with a cultivation area between 0.4 and 2 ha used to give farm-level seminars. One FF is responsible for 12–20 illustration points.
Method of engagement	Farmers can visit the points to request advice or consult with an FF.	An FF works with farmers in the fields and organizes in-field seminars to disseminate information about natural threats and disasters and to educate farmers on crop management.	An FF works with farmers in the fields and organizes in-field seminars to disseminate information about natural threats and disasters and to educate farmers on crop management.
Number of recipients (2013)	99	38	8,832

Source: From AGPPS (2013).

and drying techniques as well as guiding them on VietGAP standards and procedures for production accounting and recording. Third, AGPPS covers part of the transportation costs from the fields to the dryers. Fourth, AGPPS offers 1 month of rice storage for free. Fifth, AGPPS commits to buying rice at market prices, and sometimes exceeds that.

Expansion of FF teams since 2010-2011

As a result of LSRC, the number of FF team members has increased significantly since 2010. This is because attending training led by FFs field officers is a prerequisite for farmers' participation in the program; thus, as the program expanded, so did the FF teams. As of 2013, the FF force is active in 76 of 129 districts/towns (59%) in the Mekong River Delta. The total number of field officers (FFs) in the program reached 1,004 in 2013.

Focus on human resource development, capacity building, and scientific study for long-term targets

Recruitment process

FFs are college and university graduates who studied crop production, plantation, crop protection, biological technology, and rural development at technical schools. These field officers have a wide range of technical skills and expertise that can address the unique needs of smallholders.

Training for new employed FFs

New FFs attend a 1-month training course on management skills, market economics, and interpersonal skills, as well as on-farm teaching skills. AGPPS has organized 17 training courses for 849 FFs, who cover 24,000 ha of paddy field and work with 18,000 farmers in the Mekong Delta River Basin. The effectiveness of the FFs in supporting farmers throughout the entire cultivation process is regarded as an important driver of the expansion of AGPPS's LSRC program.

Building capacity for climate change adaptation

AGPPS, in collaboration with the Southern Plant Protection Center, Plant Protection Department, and MARD, is building FFs' capacity for climate change adaptation. For instance, 350 FFs have participated in a 1-week course on field ecology, safe usage of pesticides, natural enemies and new diseases, and environmentally friendly methods of managing diseases. AGPPS and the Southern Plant Protection Center delivered the course.[1]

Environmental protection program

In 2013, AGPPS introduced an environmental protection program led by FFs. The program provided farmers with technical assistance and guidance in the proper use of pesticides and fertilizers to protect human health and the environment. FFs help farmers to use pesticides safely and effectively by employing the four-rightness methodology,

---

1 http://www.vtvcantho.vn/CVTV/Detail/17981?id_menu=159&act=News_Detail&contr=Content& title=search.

which reminds farmers to ensure they are using the right product or pesticide at the right time, in the right dose, and in the right way. FFs also guide farmers to grow flowers around fields as an ecologically friendly method of crop protection, to employ three reduction–three increase principles, and to seed in an environmentally conscious way. Such technical assistance helps farmers reduce the cost of pesticides and fertilizers.

VIETNAMESE RICE SECTOR: EMERGING ISSUES, 2012–2015

Losing competitiveness in the global market

Vietnam is the world's second largest rice exporter, with a record high 8 million tons exported in 2013. Agriculture is considered an important cornerstone of Vietnam's economy, but it is plagued by small-scale production, inconsistent quality, lack of branding, high production costs, and volatile prices, which have been exacerbated by the global economic crisis. Since 2010, the quantity of rice exported from Vietnam has decreased continuously.

Vietnam's rice has historically been sold at lower prices than comparable rice from Thailand, India, China, and Cambodia. In 2014, Vietnam's rice exports were likely to face another difficult year, with supply outstripping demand and increasing global competition.

The decline in 2013 exports was caused by sharp drops in demand from the Philippines and Malaysia, Vietnam's traditional export markets, while Indonesia stopped importing rice altogether. A large part of the reduced demand from these countries can be attributed to boosts in domestic production in the Philippines and Indonesia. Rice exports to China, Vietnam's biggest export market, dropped as China increased its own production and the Chinese government restricted the import quota and tightened control of cross-border trade. Exports to Africa, the second largest market for Vietnamese rice, were reported to have dropped in recent years due to competition from India and Pakistan. India and Pakistan benefited from their favorable geographical locations and a competitive edge in transportation.

Small-scale farmers are in poverty: low-efficiency rice export system

Vo and Nguyen (2011) and Vo et al. (2010) divide the traditional value chain of the Vietnamese rice industry into five stages: inputs (seed, fertilizer, plant protection chemical products), production (individual farmer, farmer group), collection (middlemen), processing (husking mill, polishing mill), trade (retailer, wholesaler), and final consumption (domestic consumption, export).

The middlemen buy most of the rice from the farmers. Direct sales from farmers to food companies account for just 4.2%; husking mills account for 2.7%. Some 70.3% of the rice produced in the MRD is exported, while the remainder is consumed domestically.

Rice can be exported through one of three channels. Food companies must hold a license to export rice. The first channel is the most direct: a vertical linkage from the farmers to the food companies. This is the most efficient channel for rice producers. However, in 2010, only 4.2% of all rice produced moved through this channel. The second channel involves three middle vendors: a husking mill, a polishing mill, and then a food company. In the third channel, rice is passed through four intermediaries: a middleman, a husking mill, a polishing mill, and a food company.

In the domestic market, food companies play the role of rice wholesalers and retailers. There are also rice trading companies that focus solely on domestic markets.

The recent study "Poverty Reduction in Vietnam" by the World Bank (2013) notes that with an average cultivation area less than 2 ha per household, a rice farmer's average income is no more than US$50/month. If the cultivation area is larger than 3 ha, then a farmer's income increases to nearly US$95/month. Small-scale rice farmers can partly attribute their poverty to low levels of income from rice cultivation.

Challenges of rice export

Two decades after becoming a large-volume rice exporter, Vietnam's export-driven rice sector is still fragmented and largely unorganized. In the Mekong Delta, the country's rice granary, farmers receive little guidance regarding which varieties to grow and how to grow them. The rice grown is thus a combination of varieties, which can only be as good as its lowest-quality variety. The LSRC model in the MRD was stimulated between 2011 and 2013. About 100,000 ha, which accounts for only 5% of the land area in the region, has been organized under contract farming by agribusinesses. However, despite active cooperation with a number of private enterprises, state-owned enterprises have shown very little interest.

Vietnam's rice exports have also faced many other challenges, including inadequate transportation, poor logistical organization, use of backward processing technology, and lack of international recognition of the quality of its rice.

For many years, Thailand's enormous rice stockpile skewed global supply and put pressure on prices. Vietnam struggles to compete solely on price, and thus will need to shift its focus toward quality. Further, even as Thailand's rice sector faces challenges from its unsuccessful subsidy program that have led to larger stockpiles of rice that the government of Thailand has to pay for, there is an opportunity for Vietnam to build its brand and promote its rice.

Because of this, Vietnam is focusing on crop sector restructuring and the application of scientific and technological advances to agricultural production, prioritizing rice. The government is working on a master plan to restructure the sector by developing large rice fields, applying advanced technology in production, improving the productivity of smallholder farmers and connecting them to domestic and export markets, and increasing the value of rice and ensuring profits for farmers. Figure 13.1 illustrates the challenges highlighted above.

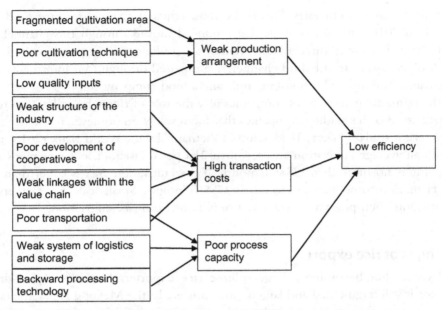

Figure 13.1 Challenges of rice export. *Source: From CAP (2012).*

LSRC program

On March 26, 2011, the MARD launched LSRC in Can Tho Province. The Vietnamese National Assembly considers this new industrial method of agricultural cultivation an important way to improve the value of Vietnamese agricultural products as well as to pursue sustainable development. It is expected that LSRC will improve the value and supply chain of the Vietnamese rice industry.

The Department of Plantation (within the MARD) noted that in 2012 LSRC was implemented in 12 of the 13 provinces in the MRD, one southeastern province, and 14 northern provinces. The total area under LSRC in the 2012 summer–autumn season reached 26,000 ha. In 2013, there were 100,000 ha in the MRD under LSRC, and it was expected that this would increase to 400,000 or 450,000 ha by the end of 2014. It is reported that LSRC helps rice producers cut production costs and improve productivity and quality.

Participating farmers have cited many benefits of the LSRC program. These include higher-quality inputs, such as pesticides and fertilizer, at lower cost. In addition, CAP (2012) reported that farmers appreciated the other benefits of LSRC, including lower labor requirements and costs, stable revenue, higher-quality yield, better storage, better input materials, and advanced technology. The farmers also added that poor farmers benefit from social development activities (a component of the LSRC) such as free health care services.

As noted earlier, AGPPS provides five different types of support to its farmers: inputs provided at cost and with 0% interest loans; on-site technical support from FFs; partial coverage of transportation costs; 1 month of rice storage for free; and commitment to buying, at minimum, at market prices. The central goal of AGPPS's value chain strategy is to create and strengthen direct linkages with farmers. With the intention to improve Vietnam's food industry, AGPPS developed a project to participate in the whole value chain of the industry. It plans to increase the commercial value of Vietnamese rice, improve farmers' income and living standards, and reduce risks to, and negative effects on, farmers. The project consists of four core components.

Dinh Thanh Agricultural Research Center

In October 2012, AGPPS established Dinh Thanh Agricultural Research Center (DT ARC) in Thoai Son district, An Giang Province. This is the first privately run research center on rice and agricultural seedlings in Vietnam. The center—a collaboration with Syngenta—is well equipped and meets world-class standards. Syngenta is in charge of technical consulting, technology transfer, and strategies, while AGPPS is the project owner responsible for constructing and operating the center. DT ARC will be connected with Syngenta's regional research center system to get updates on the latest research results relevant to Vietnam's agriculture.

Two major functions of DT ARC are human resource development and scientific research. The Center is regarded as the highest-quality research facility in the Mekong River Delta. DT ARC, in collaboration with national and international institutions and universities, studies different varieties of rice and develops new varieties that produce high-yielding, high-quality rice intended for high-end export markets. DT ARC is considered AGPPS's effort to develop a circular rice supply chain, from R&D to export markets.

The second component is the development of a specialized input material area covering 34,400 ha in An Giang, Dong Thap, and Long An Provinces. This will ensure a stable supply of quality inputs for the farmers who work with AGPPS. Additionally, AGPPS plans to develop 360,000 ha of large-scale rice fields in MRD provinces by 2018.

The third component of the project is the construction of advanced rice processing factories. AGPPS has already constructed two factories in An Giang Province (Vinh Binh and Thoai Son districts), one in Dong Thap Province (Tan Hong district), and one in Long An Province (Vinh Hung district). The total capacity of the four buildings is 400,000 tons/year. AGPPS also plans to construct another factory in Hong Dan district, Bac Lieu Province.

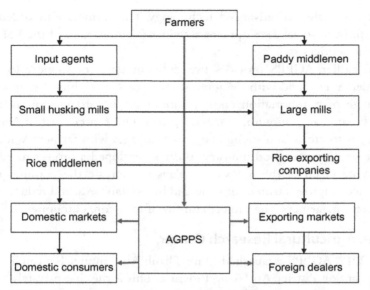

Figure 13.2 AGPPS's value chain strategy: to create direct linkage with farmers. *Source: From AGPPS (2013).*

AGPPS's market development and brand building

AGPPS' involvement in all aspects of the supply chain allows it to satisfy demanding customers as it is able to ensure a consistently high-quality supply of processed rice. The high standards and quality of rice produced through AGPPS has allowed it to expand its market to Japan, England, New Zealand, and many surrounding Asian countries. Figure 13.2 illustrates the company's strategy to reach and service smallholder farmers.

DESCRIPTION OF THE EXTENSION APPROACH

Target groups

The FF program has two target groups. The first group includes rice farmers working in fields contracted by enterprises (buyers) in the MRD. The second includes farmers who join AGPPS' technical consultation places, illustration models, and illustration points. The Toward Farmers program focuses on the MRD and some provinces in the northern, central, and southeastern regions, the Central Highlands, and Cambodia. The number of farmers involved in the program has increased at an average annual growth rate of 78% between 2007 and 2013.

Table 13.2 Partners of Toward Farmers program of AGPPS

Partners	Main roles
Central public sector	Support in terms of orientation, policies
Local government	Support the overall activities of the programs
Research institutions	Technical transfer to farmers, undertaking applied research to help farmers combat new plant diseases
Farmers' organizations	Information bridge, support to transfer information
Private companies (e.g., input producers, input dealers, agribusinesses)	To identify linkages and promote coordination with private-sector firms to strengthen the rice value chain at all levels

Source: From AGPPS (2013).

Partners

AGPPS has been developing a diverse network of partners, including entities within the central public sector for policy support, research institutions to learn about knowledge and advanced technologies, local governments for political support, farmers' organizations to facilitate communication and collaboration with farmers, and suppliers to provide farmers with fertilizers and machinery that AGPPS's associate companies do not supply.

AGPPS' partner in organizing technical training for farmers and FFs is the Southern Plant Protection Center, Plant Protection Department, MARD. Together they have implemented a series of environmental protection activities in 13 provinces in the MRD, including training courses for farmers, as well as leaders and officers in wards and districts, on safe and effective use of crop protection chemicals (CPCs), how to collect and treat agricultural waste, environmental sanitation, and potential pest problems resulting from climate change (Table 13.2).

Program content

The Toward Farmers program has three main components: (1) "Working in the field with farmers," which provides technical advice; (2) "Taking care of health together with farmers," which provides health services to the farmers; and (3) "Have fun with farmers," which is intended to enhance the social and spiritual aspects of the farmers' lives by facilitating or coordinating cultural and sporting activities.

The Toward Farmers program has a number of areas of emphasis. Major focus areas include closing technology, management, and information gaps; natural resource management; input supply; output marketing; and collecting data to improve future programming. Minor focus areas include providing management and input information, credit services, and quality assurance activities.

Program approach and design

The program was initiated by the board of managers. In 2000, after 3 years of operation, an operations manual was developed that specifies the functions of each partner, staff member, lead farmer, and participating farmer. The manual is continually updated to meet the partners' needs. Farmers are well informed about their collaboration agreements and receive incentives to follow the program's recommended cultivation procedures.

The new technologies and approaches that have been employed in the program are those that have been published. AGPPS' technical team and the FFs adapt the technologies and processes to the local conditions. In addition to its 38 illustration models and 8,332 illustration points in the MRD, AGPPS is conducting in-field experiments in which farmers can participate as well, thereby learning experimental procedures and new techniques. It also selects, from the 9,600 farmers participating in AGPPS's LSRC, 1,109 lead farmers who play a key role in bridging farmers' needs with the FF program. As of 2013, there are about 11,000 farmers enrolled in AGPPS's extension programs. The farmers are divided into groups of 10. The Coordination Committee in charge of "Go to the field with farmers" establishes the groups and coordinates their activities.

AGPPS also facilitates public–private partnerships for its extension program. Every year, the committee, in collaboration with provincial People's Committees, organizes an annual conference to review the year's performance and discuss the next year's action plan. At the conferences, farmers' groups discuss the latest issues related to technology, production, marketing, and input supplies in order to find appropriate solutions.

Value chain approach

The AGPPS's extension programs has three elements: (1) an agricultural innovation system (large-scale rice field); (2) pluralism of service providers (input provider and credit services, post-harvest supporting service, etc.); and (3) demand-driven extension services with bottom-up participatory approaches.

The Toward Farmers program focuses on technology gaps, pest control, natural resource management, marketing, input supply (mostly seeds), crop protection products, free paddy drying, rice depositories, input subsidies to eligible beneficiaries, crop and administrative data collection, consolidated reports, and market information. Figure 13.3 illustrates the structure of the large-scale rice model (LSRM) at various levels of the rice value chain.

Input supply

AGPPS directly provides farmers with high-quality inputs in a timely manner to ensure their effective use. At the beginning of the season, program officers work with groups of farmers to plan the quantities of inputs they need and when they will be needed. The company then delivers the inputs using a standardized procedure. Farmers

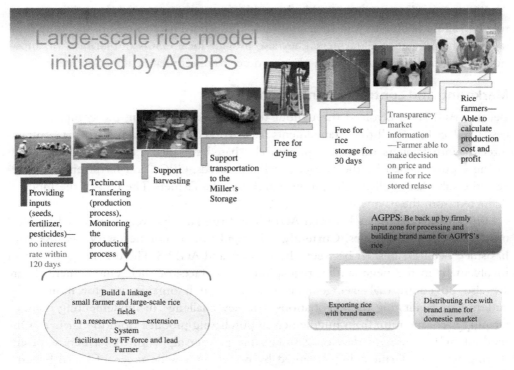

Figure 13.3 Value chain approach of Toward Farmers program of AGPPS in developing its LSRM. *Source: From AGPPS (2013).*

are advised on technical solutions for agricultural diseases. Technical advice is given according to field conditions, weather forecasts, and related information. Farmers also receive appropriate inputs at different stages of cultivation.

Program officers often visit farmers in the field during the crop season in order to get the most up-to-date information about diseases and their possible causes. This increases the linkages between technical and local information pathways, and expands the knowledge base for all. In case of disease outbreaks, the officers guide the farmers in the proper use of pesticides and CPCs. During the crop season, the officers visit the fields about 40 times and make numerous phone calls to farmers, reminding them to stay on top of what is happening in the field. The purpose is to always be responsive to farmers' problems and issues.

The program provides farmers with four types of information: crop information, disease information, crop protection, and disease remedy. The information provides specific guidance on how to use chemical products properly. Information provided to farmers reflects best practices from the field as well as inputs from AGPPS technical experts.

Extension officers connect with the public extension system and coordinate their role by calling people to join their program. They coordinate to ensure the same messages are being sent to all farmers. They also provide local CSOs and farmers' organizations with both master and action plans in order to disseminate cultivation messages.

Marketing and contract arrangements

Because AGPPS is involved throughout the whole supply chain, the company's successes reflect how well the arrangements have worked. AGPPS signs farming contracts directly with farmers through a provincially based subsidiary. This subsidiary invests in the inputs needed by the farmers, including seeds, CPCs, and fertilizer. After harvest, the subsidiary collects the produce and pays the farmers. The cost of the inputs is deducted from the payment.

Previously, agreements between AGPPS and the farmers were monitored by local crop protection authorities. Currently, this regulation is not necessary because trust has since been established between the farmers and AGPPS. There is no government involvement in the negotiations, regulation of contracts, or their enforcement. There are also no contractual arrangements with a research organization for monitoring, auditing, or facilitating the negotiations. The key challenge in implementing contract farming is competition from middlemen in purchasing produce from the farmers. One method AGPPS uses to develop a sustainable partnership with its farmers is profit-sharing with the farmers, implemented by providing a wide range of social benefits, including, health care services, cultural events, and equity offering to farmers.

AGPPS's Toward Farmers program has been self-financed. For its first program, "Working in the field with farmers," the majority of the financial resources are for wages. For the second program, "Taking care of health together with farmers," AGPPS established a social fund, which is funded from the company's after-tax profit. In 2013, AGPPS created a specific fund for science and technology development for its R&D activities within the Toward Farmers program.[2]

In order to strengthen cooperation with local partners, AGPPS issues new shares for employees, agents, and lead farmers. The lead farmers hold the most responsibility of those farmers involved in the Toward Farmers program. In 2013, there were about 1,109 lead farmers among the 9,600 farmers participating in AGPPS's LSRC. The lead farmers receive priority in becoming shareholders of AGPPS. The first round of shares issued to farmers went at about US$1.50/share.

Lead farmers qualify for the program by meeting the following six criteria;[3] they must: know and have applied technical solutions and innovations; be able to teach

2 http://quocthien.net/angiang/index.php/dai-hoi-co-dong-AGPPSs-2013-thanh-cong-ngoai-mong-doi/?lang=en.

3 http://sgtt.vn/Kinh-te/167541/Nong-dan-canh-dong-mau-nhieu-co-hoi-tro-thanh-co-dong.html.

other farmers; commit to not withholding any information from other farmers; be able to comprehensively track costs and revenues in order to conduct cost–benefit analyses; collaborate with FFs to take care of vulnerable families/households; and be able to disseminate health care knowledge to the local community.

ANALYSIS OF THE PRIVATE EXTENSION WITH RESPECT TO RELEVANCE, EFFICIENCY, EFFECTIVENESS, EQUITY, SUSTAINABILITY, AND IMPACT

Perspectives of the producers

Farmers' participation in AGPPS's extension programs—such as "Working in the field with farmers" and LSRCs—is voluntary. The programs' activities are designed to meet field-level needs, thus making the program relevant. The program aims to provide the right solution at the right time in the right place. Providing the right solution means ensuring that farmers receive sufficient resources (technical assistance, CPC products, and financial resources) to benefit from large-scale cultivation. In terms of timing, LSRC was initiated when farmers were facing terrible diseases and so were eager to join the program. The MRD was considered to be the right place because farmers used to cultivate large fields. Thus, farmers are somewhat familiar with LSRC. This situation is different from northern Vietnam, where farmers own small, fragmented pieces of land. LSRC farmers reported significant profit increases, making the program efficient in meeting their information and production needs. By joining LSRC, farmers are able to reduce uncertainties resulting from disease and price fluctuations for both inputs and outputs. Farmers are able to see the benefits of LSRC, especially higher-quality of rice, less disease, less harvest loss, lower production costs, and less labor needed. LSRC farmers are able to become shareholders in AGPPS. LSRC farmers are able to access market information and make decisions (about buying input materials and selling rice) based on their own cost-benefit analysis. Following safety precautions and guidelines for environmentally friendly usage of pesticides and fertilizers improves farmers' health. Higher incomes provide farmers with the opportunity to improve their living conditions, such as better health care and social services, as well as higher education.

Perspectives of the extension workers

The LSRC and FF program provide college and university graduates, who often lack work experience, with job opportunities and field experience. AGPPS's capacity-building courses provide extension workers not only with interpersonal skills (for better collaboration with farmers) and best field practices, but also with new knowledge. The AGPPS programs use a variety of methods and strategies to reach the intended farmers. This increases the range of extension and advisory services they can provide, thereby increasing their effectiveness at meeting farmers' varied needs. AGPPS programs run efficiently by training additional FFs to serve farmers in their own communities. This increases knowledge and information flow, which increases benefits to all participants. Extension workers can become shareholders in the business. Extension workers are able to access information about all aspects of the business. Therefore, they are able to make their own judgment about the costs and benefits of becoming shareholders. LSRC offers extension workers professional careers. The best performers can become field experts and/or researchers at DT ARC. The MRD is developing a comparative advantage in LSRC by training its farmers with its high-quality and experienced extension workers. Payments to extension workers are higher since business performance is improved. The program provides the right solution, while LSRC provides job opportunities and utilizes the labor force of young graduates.

Perspectives of company management

LSRC provides AGPPS—the largest crop protection supplier in Vietnam—with the opportunity to develop a closed supply chain for the rice industry. AGPPS is able to manage the whole high-quality rice production process, maintain close collaborative relationships with farmers, and develop a skilled extension workforce. The program prides itself on providing the "right solution," by securing a stable supply of high-quality rice and investing in large-scale production to benefit from economies of scale. The company also sells agricultural input materials (fertilizer, pesticides, and related materials), which are provided to farmers as advance payment. The program also claims to have been initiated at the "right time": LSRC was introduced when the government decided to restructure the Vietnamese economy to curb economic turmoil and reform the country's agricultural sector. Therefore, in addition to addressing farmers' needs, AGPPS receives strong policy support from the country's political leaders. AGPPS has benefited from improved business performance and financial profit. With an extension workforce that works directly with farmers, AGPPS has the advantage of having interdependent relationships with its clients who buy crop protection products and other agricultural input materials. In doing so, AGPPS has secured a stable supply of high-quality rice. Large-scale business allows AGPPS to invest in R&D activities, for example, in the establishment of DT ARC, and in improved collaboration with international partners such as Syngenta Group. LSRC requires substantial investment to reach a critical mass of production. Fortunately, AGPPS's business is thus far doing well and is strongly supported by political leaders. In light of this, AGPPS has created a corporate Science–Technology Development Fund for R&D activities.

Given the clear benefits to stakeholders, AGPPS has become a notable partner. It offers ownership to farmers, extension workers, and other stakeholders in its LSRC. Strong business performance has allowed AGPPS to improve local infrastructure and offer better social services to the communities in which it works. The success of LSRC boosted local economic growth. Due to its successful LSRC, AGPPS has been designated as the agency to implement the national program of improving the value of, and developing the brand of, Vietnamese rice. AGPPS was the first food corporation to offer ownership to farmers. This may create an industrial norm of more equitable distribution of profit. AGPPS is the first privately run corporation to efficiently participate in the National Program on Training and Creating Jobs for Rural Labor through its rice cultivation training program at the DT ARC.

Perspectives of the community at large

AGPPS provides local communities with better infrastructure and social services, including health care and education. It provides farmers not only with knowledge about agricultural economics and cultivation techniques, but also with market information. LSRC helps local communities improve their living standards by excellent business performance resulting from collaboration between community members and the other players. By training local farmers to be leaders in their community, AGPPS's programs use their financial and human resources efficiently. Farmers and other members of the local communities are at the center of LSRC, and the production process is based on interdependent relationships between the farmers and AGPPS. The program stimulated a change in the way of thinking about cultivation methods through the "seeing then believing" approach.

COMPARISON WITH PUBLIC EXTENSION AND OTHER PRIVATE SECTORS IN VIETNAM

AGPPS is regarded as the best extension service provider in Vietnam. It has created a closed-circuit supply chain of rice production, providing farmers with technical assistance throughout the production process. AGPPS makes farmers a real part of the business by offering ownership. AGPPS also benefits from strong policy support. There are numerous differences in the structure and strategies of AGPPS compared to the public system in Vietnam.

Like other private companies that use the value chain approach (including ITA Rice Company, ADC, GENTRACO), AGPPS links with farming households to (1) train all farmers in the optimal technology to grow the rice varieties needed for high-quality domestic consumption or for export; (2) supply the farmers with the inputs they need to produce raw product at the lowest production cost but highest quality; and (3) buy the product from the farmers at a premium price, 10% above the current market price. However, AGPPS offers more unique services that could allow farmers to participate actively as members of the company by providing them the chance to become shareholders and receive health and social services.

The most remarkable difference between private extension and public extension is that most private extension operates through contracts with farmers, driven by the value chain approach. By comparison, public extension generally uses a top-down approach: State extension agencies often have few resources for training and only train on production practices. This leaves large information gaps regarding post-harvest practices such as storage and marketing. The private sector provides tailor-made training based on market need, and at all the stages of the supply chain, to ensure the outputs generated will meet the quality and quantity demanded by the market.

Another difference is the ways in which public and private extension systems develop and manage human resources. Private extension providers self-fund the incomes of the technical workers, while state extension officers are paid by the state. Because of this, the private extension system is incentivized to monitor and evaluate the efficiency and productivity of its extension workers, while the public extension system does not.

Additionally, the capacity for communication, one of the most important tools in extension, varies greatly between the public and private extension approaches. In Vietnam, the public extension system has favorable support from public communication channels such as television and newspapers. It also receives substantial financial resources annually from the state budget to execute their work. In contrast, the private extension system cannot access the national communication channels without cash payments. The reason is that communication agencies view information from the private sector as marketing or as a public relations tactic. AGPPS is a rare player among private extension providers in that it can afford to mobilize mass media as well as ICT in their extension services.

The last difference between public and private extension in Vietnam is that public extension often facilitates the involvement of a third party to provide quality standard training and certification for a production system, while the private system may not be willing to invest in that. However, it may be willing to support farmers to follow non-certified quality management methods. This is a potential area in which to develop a public–private partnership.

UNIQUENESS OF THE MODEL AND VALUE ADDED

AGPPS is the largest manufacturer and distributor of crop protection products in Vietnam. Founded in 1993, AGPPS was privatized in 2004. AGPPS has extensive distribution networks, flexible sales policies, and a strong base of loyal customers—in part because it offers direct technical assistance to farmers. Its volume of sales has increased in the last 5 years, with a cumulative annual growth rate of 27%.

As the largest distributor for Syngenta, AGPPS offers a comprehensive supply of high-quality products, provides advanced technical assistance to customers, and has strong marketing campaigns. AGPPS continues to focus on its core competencies and retaining its competitiveness in the plant protection segment of the market. However, it is also expanding into seed supply, and product collection, processing, and storage, so as to provide comprehensive crop solutions at any stage of production. This entails diversifying its business into related products and services.

In order to research and develop high-yield rice varieties that resist pests and diseases, as well as to produce clean and high-quality rice that meets both local safety standards and the demands of exporters, AGPPS has started a number of large projects. For instance, AGPPS established an advanced agricultural research and training center in Vietnam, which, in collaboration with national and regional institutes and universities, will develop and provide rice growers with comprehensive solutions. The company also invested in large-scale rice fields in several provinces in the MRD, the country's rice bowl. The total investment is reportedly nearing $146.5 million, of which $20 million is government funding.[4]

LESSONS FOR REPLICATION AND SCALING UP

Given that the LSRC models have been implemented in a large region facing unstable weather conditions (especially due to climate change and upstream dams) and complicated diseases, the FFs need a vast amount of crop management knowledge. The quality of knowledge transferred from FFs to farmers and the efficiency with which it is done are prerequisites for high-quality outputs, which are needed to establish Vietnamese rice as a brand. This helps the exporters to secure revenue and the farmers to enjoy stable incomes. AGPPS plans to increase the number of FFs to 4,000, about four times the existing number. To this end, there is a great need for collaboration with scientists and researchers to train the new FFs. Replication and scaling up will require expansion of the capacity of extension personnel at all levels.

To deal with these problems, the Toward Farmers program is designed with a full package of technical assistance, technology transfer, health care services, and various social benefits to build loyalty to AGPPS. Moreover, AGPPS is developing a plan to improve the company's engagement with farmers' communities through profit-sharing schemes. A key lesson for further scaling up is the development of trust between the farmers and AGPPS. This is a long-term process.

CONCLUSIONS

AGPPS has invested in strengthening the rice supply chain by providing farmers with technical assistance throughout the production process. AGPPS enables farmers to take on a business perspective on their production by offering ownership opportunities.

---

4 http://english.thesaigontimes.vn/Home/business/other/27261/; http://www.baocantho.com.vn/?mod=detnews&catid=72&id=123831.

The company also takes advantage of its financial strength and strong policy support. AGPPS's model of investment in farmers is considered a novel method and an important solution to improving the value of Vietnamese rice products and to achieving sustainable development. The company offers a number of programs and projects, as outlined above, to meet the unique needs of farmers. This also allows the company to adapt and expand its extension and advisory services as the market and agribusiness environments change.

While developing extension programs, AGPPS improved its collaborative methods, aiming to add more value to these programs through specific engagement with farmers with, for example, preferential stock offerings. A proper profit sharing scheme will not only strengthen the farmer–exporter link but will also further incentivize a high quantity and quality of production. Such improvements are necessary for AGPPS to be confident in investing fully in LSRC throughout the MDR in the next 5-year period (2013–2018).

Early successes have made AGPPS a reliable brand and partner of both local farmers and government. AGPPS has received strong political support from central and municipal leaders, especially in 2012 when AGPPS was designated the implementing agency of the National Program on Developing Brands and Improving Quality of Vietnamese Rice in 2013–2018. The program provides AGPPS with financial support as well as international connections and collaboration, and strengthens AGPPS's collaboration with the scientific community and provincial governments.

In addition to these advantages, AGPPS possesses four strengths: stable finances, high-quality products, a reputable brand based on its Toward Farmers business strategy, and a cadre of young, well-educated agricultural engineers. It is expected that AGPPS's success story will boost the Vietnamese rice value chain while enhancing the livelihoods of the farmers with whom it works.

REFERENCES

AGPPS, 2013. Sustainable development strategy of AGPPS vision to 2020.
CAP, 2012. PRA exercise to farming household at Vinh Binh, An Giang.
Vo, T.T.L., Bush, S.R., Sinh, L.X., Khiem, N.T., 2010. High and low value fish chains in the Mekong Delta: Challenges for livelihoods and governance. Environ Dev Sustain 12, 889–908.
Vo, T.T.L., Nguyen, P.S., 2011. Value chain analysis of rice product in the Mekong Delta. Scientific Journal of Can Tho University 19a; 96–108.
World Bank, 2013. Poverty Reduction in Vietnam: Remarkable Progress, Emerging Challenges. Available at: <http://www.worldbank.org/en/news/feature/2013/01/24/poverty-reduction-in-vietnam-remarkable-progress-emerging-challenges>.

CHAPTER 14

Private Sector Extension—Synthesis of the Case Studies

Suresh Chandra Babu[1] and Yuan Zhou[2]

[1]International Food Policy Research Institute (IFPRI), Washington, DC, USA
[2]Syngenta Foundation for Sustainable Agriculture, Basel, Switzerland

Contents

Agricultural development remains a major determinant of the economic growth of many developing countries. It also remains an important source of employment for millions of rural households whose livelihoods are derived from agriculture. In order for the agricultural sector to grow, its total factor productivity must increase. Knowledge of innovative ways of producing crops, raising livestock, and fishing contributes to an increase in total factor productivity. Thus, access to knowledge is a key

Knowledge Driven Development.
DOI: http://dx.doi.org/10.1016/B978-0-12-802231-3.00014-0

253

determinant of agricultural growth. However, the system of agricultural extension, the traditional channel linking research and farmers to raise issues and disseminate knowledge, has broken down in many developing countries.

In the wake of the recent food crisis, developing countries and developed countries alike have placed more emphasis on increasing the productivity of their agricultural sector. There is increased interest among most of the developing countries in reforming their public agricultural extension systems. Agricultural extension and delivery systems have become increasingly pluralistic to accommodate the decreasing public-sector involvement; a number of different actors and players participate in the knowledge-sharing process to meet the specific information and knowledge needs of the farmers. Farmer-based organizations and NGOs have emerged to meet the needs of smallholder farmers, while the private sector has developed strategies to supply farmers with quality seeds and other inputs, including credit to buy the inputs and to help farmers to aggregate their produce.

With the increased involvement of the private sector in extension and rural advisory services, there is a need to understand the role they play, the gaps they fill, the challenges they face, and the benefits and costs of their involvement. It is also important to identify the role of public policy in enhancing public–private partnerships and opportunities to integrate the services provided by various entities in order to increase the farm gate value of agricultural products.

Discussion around the private provision of extension and agricultural services is not new. The decline of public extension in several developed countries over the last century has resulted in the introduction of various forms of private extension approaches, where farmers can get access to private extension workers and consultants through some form of public support. However, the role of the private sector in delivering extension and advisory services has increased through the direct participation of the input suppliers. Although their prime motivation is selling their chemical inputs, fertilizer, and machinery to the farmers, these private companies have increased their investment in the marketing and sales force which also perform extension services. This system of private extension also largely applies to the developing countries wherever the multinationals have been able to operate alongside the public extension systems. In fact in many developing countries, with the decline of the public extension, the input dealers and the marketing agents of the chemical companies became the most relied-upon source of extension service. This is particularly the case with cash crops such as cotton, cocoa, and other high-value crops in developing countries. More recently, with the advent of contract farming and value chain development, where output aggregators have emerged as a source of extension, private extension has taken on a new dimension of its own, combining all aspects of private extension. Accordingly, cost recovery for extension services has also evolved to sustain such efforts by the private sector. Yet little is known about this emerging approach to extension provision by private-sector entities in developing countries. In this book we have synthesized ten case studies from seven countries to highlight a variety of approaches of the private provision of extension services.

The approach used in the case studies presented in the previous chapters closely followed guidelines developed for such purposes by the Global Forum for Rural Advisory Services, adapted as necessary to address the various country contexts. The case studies attempted to understand the relevance of private extension programs and the factors that contributed to the emergence of such programs. Initial attempts to identify cases that specifically addressed output aggregation and input supply separately were not successful, and most of the private-sector operations invariably involved both operations, albeit to varying degrees. Output aggregation and providing farmers with a market outlet were the primary objectives of most private companies engaging in extension.

VALUE CHAINS AND PRIVATE EXTENSION

The private extension case studies analyzed in this volume look at commodities around which value chains have been formed. Studying the value chains themselves is beyond the scope of this report. Here, we focus on the extension and advisory services provided at the early stages of the value chains. The approach to organizing farmers to provide these services is generally characterized by a contractual arrangement between private-sector companies and the farmers they service. While the contracts were primarily meant to ensure that the outputs sold to the company would meet specific standards, provision of inputs, assistance in accessing credit, and the availability of extension advice are also laid out in the contracts. Provision of quantifiable inputs such as seeds, seedlings, fertilizer, chemicals, and irrigation systems are explicitly stated in the contract and are often repaid by the farmers on a cost-recovery basis during the buying back of the outputs. However, the cost of extension advice is not explicitly charged to farmers in the cases studied; in the case of contract farming, it is incorporated into the price the private companies pay for the farmers' commodities. Due to low levels of monitoring, it remains a challenge to quantify the exact cost per farmer of the extension services provided. It is clear, however, that it is the farmers who share the cost of extension offered by the private companies, in most cases indirectly.

Actors in the value chains

For all the commodity value chains studied, the actors included the farmers, input suppliers, wholesale traders, retailers, consumers, and a regulatory authority, such as the international certification agency for organic farming. In most of the case studies, the entry of the private sector to buy outputs on a contractual basis resulted in a reduction in the role played by the existing traditional output buyers. In Brazil, Rio de Una's presence reduced the role of wholesalers and local traditional markets in the vegetable sector, although these entities remain in existence for traditional, non-contracted farmers. Additionally, some of the contract farmers continue to sell their excess produce through the middlemen. In the sugarcane industry in India, farmers had always sold directly to the local cane processing company. The emergence of contract farming helped the company in the case study to expand its coverage and increase the volume of production. It became more responsive to farmers' needs when it was discovered that other crops were potentially more profitable for farmers. In the case of the Nigerian cocoa industry, the role of the middlemen was reduced to a larger extent upon the entry of the processing company and the appointment of lead farmers. In Maharashtra, the traditional middlemen and traders in the onion industry were sidelined when the company began purchasing produce directly from contracted farmers and processing it for export markets. However, produce that did not meet export quality standards was still handled by local traders. A similar trend was observed in the vegetable value chain in Kenya and organic basmati production in Jammu, India.

Relevance of the private extension approaches to the farming community

In five of the case studies, the extension system was driven by the need to improve the quality of outputs bought by the private sector. Organizing and contracting farmers to purchase their produce in aggregate made it easier to provide farmers with extension to improve the quality of their produce because of the already established communication channels. For the private companies to remain competitive in their industry, their product needs to have a distinguishable level of quality. For example, in the case of Rio de Una, a Brazilian vegetable processor that supplies produce to such chains as McDonalds and Walmart, quality control is paramount. The nature of the processing requires that the company's farmers meet certain quality standards. For example, the organic certification needed for Rio de Una's products requires that the company educate its farmers on organic farming methods. This was also the case for organic basmati rice production in Jammu, India. In the case of KHE in Kenya, the food quality standards, such as GlobalGAP required by European customers, imply that vegetable farmers must follow a strict planting procedure and use only recommended inputs.

In the case of sugarcane in Tamil Nadu, India, it is a greater quantity of produce that is needed by the company to stay competitive, as their profitability comes from keeping their processing plant running regularly. Thus there is a need to increase the productivity of the sugarcane farmers through provision of an extension system. Until the last decade, the company relied on the public extension system to educate its farmers. However, several emerging factors such as water scarcity, non-farm opportunities, and government subsidy of perennial crops such as cashew have reduced the acreage under sugarcane. To attract new farmers into sugarcane cultivation to meet the cane supply needed by the crushing factory, the company identified information gaps among the farmers and began a system of highly relevant advisory services.

As illustrated by Jain Irrigation Systems Limited's specialization in onion cultivation in Maharashtra, India, private companies have also identified niche markets in which to invest. They use their large presence in the region effectively to enter into, and remain in, a contract with farmers. Onions are in strong demand throughout the year as Indian cuisine uses onions on a daily basis. Onions also yield a good return on the international market, and India has exported onions to Sri Lanka, Bangladesh, and a number of Middle Eastern countries. Onion production requires much irrigation, which gives the company an added edge as they are the leading manufacturer of micro-irrigation systems in India. Thus the relevance of the extension advice is inherent in the choice of the commodity value chain and the need to increase the quality and/or quantity of outputs to make the value chain profitable.

In the case of Frijol Nica, the extension system was driven by input suppliers to enhance bean productivity through adoption of quality inputs. The program was specifically designed to address constraints such as the lack of finance and technical assistance faced by bean growers in Nicaragua. Advanced inputs such as certified seeds and crop protection products, as well as auxiliary advisory services, were provided to boost productivity and profitability. Given the large number of bean farmers in poor areas, such a breakthrough in productivity gains is highly relevant for ensuring food security and reducing poverty. The higher price paid by Sarveshwar Foods for organic rice production in Jammu also helped to increase the income of the farmers, thereby improving their welfare.

Effectiveness of the private extension system

Another indicator of a successful extension system is its effectiveness in reaching its goal. By design, the private companies have been effective in meeting their extension goals as they have been able to become competitive suppliers in the global market. For the most part, an ineffective extension system would result in the company being unable to meet its target output levels. In Brazil, Rio de Una's extension workers are on call every day and respond to the contract farmers needs over the phone. They also operate an extension center, which receives calls from farmers 24/7 and responds to these calls within 24 h. The call file is not closed until the problem is addressed to the satisfaction of the farmer. In addition, the extension training center is open 7 days a week so farmers can stop by and learn the next steps in crop production and/or have questions addressed in person. Jain's extension program for onion farmers has a ratio of extension workers to farmers as high as 1:50, and the extension worker's coverage is limited to 100 acres. This increases the intensity of the services provided to farmers. Thus, the case studies collectively indicate that the private sector has been effective in providing the extension services to the farmers they serve.

Efficiency of private extension

Are private companies efficient in their delivery of extension? The companies are explicit in charging for the inputs they supply to the farmers at the beginning of the season. These costs are fully recovered when the produce is purchased from the farmers by the company. In addition, the price the companies pay for the farmers' produce reflects the costs they incur in providing technical advice to the farmers. Further, since the companies are typically the only outlet for the farmers to deliver their commodities, there is an element of monopsony at work, over which the farmers have little control. In the case of Rio de Una, farmers are expected to meet their contractual obligations before they sell to the open market. In India, sugarcane farmers are geographically constrained to sell their cane to their local sugar company, which uses its monopsony power to recover the full cost of their extension operation. Similarly, Jain's onion growers sell exclusively to the company for processing for export.

Sustainability of the private extension system

The sustainability of the private extension programs depends on the sustainability of the private company itself and its ability to maintain its competitive edge and hence its market share. As long as the company continues to profit from its contract farming operations and is able to offer an attractive price to its farmers, the private extension system is sustainable. The shared-value addition is thus a key to the sustainability of the private extension service. For input suppliers, as illustrated in the Frijol Nica program, the sustainability of private extension relies on the functioning of the whole value chain and value addition from all participating partners, as well as innovation in financial arrangements for access to credit.

Impact of the private extension system

Assessing the impact of the extension system is a complex task due to the various factors affecting farmers' production and market decisions. Here we present an abbreviated assessment of the impact of the private companies studied and give directions and relative sizes of impact with respect to five indicators. This is given in the format of "before and after," keeping before at the "++" level. Private extension has been beneficial in terms of yield increases; however, there are mixed results in terms of cost reduction. The consistency of the quality of produce delivered by the farmers improved significantly. The net profit of the farmers also improved significantly, as did natural resource management (Table 14.1).

Feedback on the private extension services from various perspectives

The case studies also provided feedback from the contract farmers themselves, the extension workers, the communities in which the companies operate, and company

Table 14.1 Summary of benefits from private extension

Benefits of private extension	India/ sugarcane	Kenya/ vegetables	Brazil/ vegetables	Nicaragua/ beans	Nigeria/ cocoa	India/ onions	India/ oilseeds	Bangladesh/ vegetables	India/ organic rice	Vietnam/ rice
Yield increases	+++	++++	++++	+++++	++++	++++	+++	++++	+++	+++
Cost reductions	++++	+++	++	++	++	+++	++	+++	++	+++
Quality and consistency of outputs	+++	++++	++++	+++	+++	+++	+++	++++	+++++	+++
Increase in net profit	+++	++++	+++	+++	++++	++++	+++	++++	++++	+++
Natural resources benefits	++++	+++	++++	++	+	+++	+++	+++	+++	+++

Source: Author's compilation.

management. The responses were mainly positive. In the absence of a well-functioning public extension system, any intervention that improves farmers' access to information and quality inputs is likely to be hailed as a positive development. Yet respondents also identified several issues in the delivery of private extension.

Farmers in all the case studies noted positive outcomes from the intervention of the private sector. Their ability to produce crops in a more organized cultivation system with regular advice from extension workers was the main benefit for the participating farmers. Table 14.2 summarizes the feedback from the farmers with regard to various stages of crop production.

The responses of the farmers' communities were generally positive in all of the cases studied. However, they varied slightly depending on who was consulted in the community. Public extension workers felt that the farmers were exploited in terms of the prices they were offered by the private sector, even though the extension services helped farmers overcome various obstacles throughout production. For example, while the sugarcane farmers expressed general satisfaction with the extension services provided by the Parry Company in southern India, community leaders felt that the company was not flexible in their harvest timing and that the prices were fixed without much farmer consultation. They also expressed dissatisfaction with the way the company paid for the extension services they provided, namely by holding back part of the payment from the farmers. Among the Brazilian farmers' communities served by Rio de Una, there was a more positive reaction. However, the farmers who did not participate in the program did not do so because they felt that the contract was too restrictive.

Feedback from the extension workers employed by the companies was positive. In almost all of the cases, they pointed out the poor functioning of the public extension system and that their companies' involvement is filling those gaps. They claimed the services they provided integrated production advice with supply of inputs. They felt that their intervention largely reduced the uncertainty in the quality of inputs they were using. Having an assured and timely market by preparing farmers to meet quality standards was seen as the most important benefit of the private extension. Extension workers felt a high level of job satisfaction from addressing farmers' challenges on a regular basis, and they are seen by the farmers as reliable. They also felt that the capacity developed by the lead farmers would have long-term impacts as the farmers can continue to serve the farming community even in the absence of a contract or in the event that the farmer ceases to participate in the program.

Feedback from the company managers was invariably positive. They saw the approach as a win–win proposition for both the company and the farming communities they serve. They saw the role of the company as helping farmers to attain a higher standard of living through engaging them in the value chain. However, company managers were reluctant to have their extension workers go beyond the extension services to address the social and personal challenges of the farmers.

Table 14.2 Summary of the qualitative feedback from farmers on the extension service compared to public extension (ranking: +++++, excellent; ++++, very good; +++, good; ++, fair; and+, poor)

Crop/nature of advice	India/ sugarcane	Kenya/ vegetables	Brazil/ vegetables	Nicaragua/ beans	Nigeria/ cocoa	India/ onions	India/ oilseeds	Bangladesh/ vegetables	India/ organic rice	Vietnam/ rice
Pre-planting advice	+++	+++	++++	+++	++	++++	++	+++	++++	+++
Input supply— seedlings/ seeds	++++	+++	++++	+++	++	++++	+++	++++	++++	++++
Input supply— fertilizer and chemicals	++++	++	++	++++	++++	++	++	++++	++++	++++
Help with crop loans	++++	++	+++	+++	+++	+++	++	++	++	++
Irrigation systems	++++	++	++	++	++	+++	+++	+++	+++	+++
Intermediate cultivation/ weeding	+++	+++	++	+++	++	++	+++	++	++++	+++
Harvesting help/advice	++	+++	++++	++	+++	+++	+++	+++	++++	+++
Post-harvest advice	++	++++	+++++	++	++++	++	+++++	++++	++++	++
Marketing advice/help	++	++++	+++	+++	+++++	++++	+++++	+++	++++	++++

Source: Author's compilation.

BROAD LESSONS FROM THE PRIVATE EXTENSION CASE STUDIES

Creation of shared value

The private companies studied in this volume aimed to create shared value for the farmers, input suppliers, and output aggregators. In most cases, the companies also acted as input suppliers to ensure the quality of the seeds and fertilizers used, while controlling the quality of the outputs. This approach either resulted in reduced cost for the farmers or enabled them to get better access to needed inputs at the right time for crop production, which contributed to reductions in yield losses. This further increased the net profit of the farmers and produced the outputs needed by the companies at reasonable prices. Without such shared value, this model of private extension would not be sustainable.

Provision of integrated services

A major factor contributing to the success of the private extension programs examined in these case studies is the integration of services provided to the farmers. In the past, when the private companies that provided only the chemical inputs supplied information to the farmers on pest control, for example, this was considered only partial information. Such information sharing was also seen as biased since the input dealers or the marketing agents of the chemical companies were seen as pushers of their products. This is still the case in areas where private companies are operating. This model was to some extent seen as exploitative and also resulted in competition among the private chemical companies. Whoever was able to give a better commission to the input dealer at the village level was able to sell more of their chemicals, and farmers were always at the mercy of these competing companies. In the case studies examined in this volume, the private companies working with both farmers and input suppliers have facilitated a more efficient exchange of inputs, including seeds, fertilizer, chemicals, intercultivation equipment, and micro-irrigation equipment. This, combined with the provision of quality advice at every stage of crop production, has helped farmers to get the inputs they need and resolve issues in a timely manner. In addition, building linkages with credit institutions, as in the case of the India sugarcane study, also helps farmers who may not have time to shop around for financing options to obtain a loan on their own. While the role of individual private input dealers is not fully eliminated, the aggregating companies are able to play a useful role as integrators of the input suppliers, and extension messages are delivered in a more harmonized manner.

Better research–extension linkages

One of the major benefits of the private-sector extension programs is the effective use of innovations and technologies for increasing the productivity of farmers. The private companies studied in the report have tapped into research undertaken by the public sector, or they have developed their own research programs. For example, in Brazil, Rio de Una was able to connect farmers with a seedling company that conducts its own research, investing resources in breeding new varieties. The white onion variety promoted by Jain Irrigation Systems is a product of their own plant breeding program. In India, EID Parry Company has been conducting its own research to test the varieties of sugarcane released by the public research systems, to improve the agronomic practices of the farmers. It is in the interest of the output aggregating companies to guide the farmers in their choice of varieties to meet the quantity and quality expectations of the output markets they are dealing with. In Nigeria, for example, the processing company directs the cocoa farmers who are planting new saplings by recommending cocoa varieties depending on the quality of the cocoa expected in the export market. This is possible only through close engagement between the farmers and the extension agents of the company through the contractual arrangements. Private-sector companies thus engage to varying degrees in such research communications specifically to improve farming techniques and present this research to farmers. This is in contrast to what occurs in many developing countries, where weak research–extension linkages are a serious problem in the public-sector extension system.

Market connections—assured market access

The private companies studied in this report play a major role in output aggregation. This means their contracted farmers have an assured market for their produce and the uncertainty of market access and demand is eliminated to a large extent. This is invariably the case for all of the commodities studied. In fact, output aggregation is clearly a safe entry point for the private sector to engage directly with farmers. The need for extension arises to ensure the correct quantity of produce can be delivered to the company and that it meets a minimum level of quality. Thus extension plays a highly complementary role in the value chain development of these commodities. Thus, development of the value chain for high-value crops in developing countries may have additional benefits in better engagement of the farmers for extension through the private sector. However, the downside of this is that when the demand for a commodity goes down or the private company is not managed well, the farmers may not get their extension services and may have to depend on other sources.

Reduction in price uncertainty

An additional benefit of the private-sector contract with the farmers is assured prices. The company expects a certain level of quality in the produce delivered for the price agreed upon in the beginning of the season. This requires the company to provide needed knowledge on the maintenance of produce quality during harvest and post-harvest handling of the produces. The prices are announced early in the cropping season for annual and seasonal commodities, and this for sure reduces uncertainty about the prices farmers will receive after harvest. However, there is an element of anxiety among the farmers that they may be able to get higher prices if they were to sell outside the contract. Some contracts allow this. For example, in Brazil, Rio de Una is able to allow the farmers to go outside the contract and sell in the free market as long as the farmers meet their delivery requirements. However, this is not the case in all of the case studies studied.

Inclusive innovation

In all the cases studied, the private companies tend to benefit from improving the knowledge base of the farmers. By working with them, the extension agents are able to identify challenges farmers face in various production, protection, and processing aspects of crop cultivation. This helps to engage farmers in the development of solutions. Such solutions are much more likely to be adopted by the farmers than those that are developed without consultation. Thus, through their extension agents, private companies have identified opportunities to work with the farmers to find solutions to their problems. This again forms a major aspect of the advisory services provided by the private companies. For example, in Brazil, the need to meet certification standards for organic produce forces the private company to provide extra advisory services on finding non-chemical alternatives for plant protection challenges, which has led to the development of new plant-based materials.

Enhanced use of quality inputs (seeds, fertilizer, chemicals)

Timely access to inputs and their efficient use are important in increasing farm productivity. By reducing the barriers to accessing quality inputs and providing advice on the proper use of those inputs, private companies have effectively served the farmers. In all the case studies presented here, farmers reported that their access to quality inputs was a major benefit of engaging with the private companies. The extension agents of EID Parry in India are able to develop input purchase plans for the individual farmers they contract with. This helps to ensure that the farmers get quality inputs at the time they are needed. Due to high demand for these inputs at almost the same time in a given region, it also helps the company to organize the supply of the inputs for their region as a whole. This also helps them to have stronger negotiating power with the input suppliers and thus get better prices for the farmers.

Product differentiation

The quality of outputs and the consistency with which they are produced ensure that the aggregating company can produce a differentiated product. This is particularly true for organic products produced by the companies. However, maintaining this quality requires continuous engagement with the farmers and development of the farmers' understanding of product differentiation. The private companies studied here invested in developing tools and strategies for the farmers to help them maintain quality. Such investments are normally missing in the public-sector approach to extension in developing countries.

Solidarity building

The private companies have been able to bring farmers together to contribute to the same cause and solve the common challenges they face. The lead cocoa farmers in Nigeria, for example, have come together to develop community-based initiatives that address local socioeconomic problems. Renovating village schools and maintaining local bridges have been undertaken by farmers' groups to help their communities. However, the cohesiveness with which farmers are able to organize also helps to increase their voice and power in dealing with the private companies. In the case of sugarcane farmers in India, farmers were not well organized to address common problems that they could face with the company. There is often talk among the farmers that the company could increase its payment for the cane, but this was not taken up as a cause by the farming community. This is possibly due to the fact that the company remains the only buyer of sugarcane from the farmers in the region.

Cost recovery

Who pays for extension is still a major issue which is often not discussed among the farmers. It is often argued that when information is free, farmers do not take it seriously. However, research has shown that the willingness to pay for information is low among small-scale farmers. By recovering the costs of extension implicitly via the payment made to the farmers in contract farming systems, the private extension providers have internalized the cost of extension. In the case of input supplier–driven extension systems, the costs of extension are recovered from increased sales of inputs. Currently, since the cost of extension is recovered through the prices dictated by the private companies, there is little transparency in the way the farmers are charged for information services. While the system of extension through private companies aggregating produces may be better than the private input companies whose ulterior motive is to sell their chemicals, farmers may be paying more for extension through the output aggregation system of extension. There is a need to conduct in-depth studies on cost recovery and compare with the data on farmers' willingness to pay.

Demand driven

A major criticism of public extension systems is that they are supply driven and do not meet the information needs of farmers. The extension advice provided by private companies is necessarily demand driven, directly addressing the information gaps of the contract farmers. This is partly due to the contractual arrangement that the companies have with the farmers and partly due to the nature of the extension and advisory services provided to the farmers. In most cases studied here, since the farmers grow homogeneous crops, it is much easier for extension workers to deal in a systematic way with specific issues related to the crop as the issues need to be addressed during the cropping season.

Capacity development

The private companies studied in this volume provided farmers with the capacity to produce a higher quality and quantity of produce. For example, farmer training provided by the sugarcane company in India allowed the introduction of new methods of cultivating sugarcane. This investment has long-term impacts on the capabilities of farmers and has had spillover effects, especially in enhancing production of crops that farmers are not under contract to grow. It has also influenced farmers who are not currently under contract to switch to new methods of farming sugarcane. Cocoa farmers in Nigeria, for example, learn from the lead farmers, who are continuously trained by the extension workers of the private company. Vegetable growers in Kenya learned to work with different international standards including GlobalGAP. Such mutual learning has been helpful in reinforcing and updating the farmers' knowledge base in the case study areas.

CHALLENGES AND WAYS FORWARD

Despite various positive impacts of private extension programs on farming communities, several challenges remain before the private-sector model of extension can be scaled up. First, farmers feel a need for the government to monitor the private companies so that the farmers are treated fairly. Second, a major challenge, common to all case studies, is respecting the terms of the contracts (by both parties) and solving disputes when they arise. When either party violates the contract, it is difficult to enforce a punishment due to weak justice systems. Third, while the prices paid for the physical inputs supplied to farmers are accounted for in a transparent manner, the price farmers pay for extension services is not clear, as farmers are implicitly charged by fixing the price for the commodities to account for costs of operation. This has created some mistrust of the company and could become a barrier to expanding the contract farming program. Finally, scaling up the private extension programs depends on the scaling up of contractual arrangements. However, most of the value chains run by the private companies fill niche markets, and expansion depends on the company's ability to identify new market avenues. This may require additional investment to build new processing facilities, which companies may not be willing to risk even if they are successful with the original plant. Thus, the scale (and scope) of private companies' operations may remain limited. This calls for public-sector investment to meet the extension needs of farmers who grow crops that the private companies do not process or export.

The challenges identified above could be addressed through various innovations in the way the extension systems currently operate. Some of these come out of the case studies presented here. First, there is clearly a need for better participation from the public sector to play a monitoring role, preventing or at least reducing unfair treatment of the farmers. This could be organized through public–private partnerships that could be developed by the private and public sectors. For example, in the case of Brazil, the public extension system—EMATER—has a key role to play in helping the farmers to access medium-term loans for the purchase of irrigation systems and pickup trucks, which can enhance the ability of the private company to expand in specific areas where water may be scarce or in areas where transportation to the company may be a challenge. Farmers' voices could be better heard if the farmers of the command area could be better organized in the form of farmer organizations. This would help in better negotiations of prices and services provided by the private companies. Such organizations could help farmers in other ways, bringing in local NGOs and the public sector to benefit the farmers in ways that go beyond the extension services for crop production.

A common factor contributing to the success of the private extension models described in this book is the integrated approach to bringing inputs, services, and advice to farmers at the farm gate. This requires the private company to sign contracts with input suppliers, which ensures that farmers receive quality inputs. However, it also increases the cost of cultivation. Unless these costs are covered by increased productivity, contract farming will not be sustainable. Thus, extension and advisory services become a necessary service provided by private companies to ensure their own success.

In order to evaluate the benefits of private extension services, it is important to collect data on the costs and outcomes of such services. This would help policymakers to understand the constraints in identifying and developing public–private partnerships for value chain development and related extension services.

CONCLUDING REMARKS

This chapter provided a synthesis of the major findings from 10 case studies of private extension services in seven countries, which were conducted between May 2013 and October 2013. The case studies looked at the methods of extension, the services rendered by the private companies, and the relevance of the services, as well as their effectiveness, efficiency, impact, and sustainability. Private extension programs developed and implemented as part of value chains are effective in their respective areas, but their coverage is limited. They are not a perfect substitute for public extension systems, even in the areas where they operate. One way to leverage their presence to the benefit of the rest of the farming community is to develop public–private partnerships to increase the efficiency and effectiveness of the public extension system. As value chain expansion continues in developing countries, the importance of research around experimenting with innovative extension approaches and studying their costs and benefits cannot be underestimated.

INDEX

Note: Page numbers followed by "*f*" and "*t*" refer to figures and tables, respectively.